# Springer Tracts in Mechanical Engineering

**About this Series**

Springer Tracts in Mechanical Engineering (STME) publishes the latest developments in Mechanical Engineering – quickly, informally and with high quality. The intent is to cover all the main branches of mechanical engineering, both theoretical and applied, including:

- Engineering Design
- Machinery and Machine Elements
- Mechanical structures and stress analysis
- Automotive Engineering
- Engine Technology
- Aerospace Technology and Astronautics
- Nanotechnology and Microengineering
- Control, Robotics, Mechatronics
- MEMS
- Theoretical and Applied Mechanics
- Dynamical Systems, Control
- Fluids mechanics
- Engineering Thermodynamics, Heat and Mass Transfer
- Manufacturing
- Precision engineering, Instrumentation, Measurement
- Materials Engineering
- Tribology and surface technology

Within the scopes of the series are monographs, professional books or graduate textbooks, edited volumes as well as outstanding PhD theses and books purposely devoted to support education in mechanical engineering at graduate and post-graduate levels.

More information about this series at http://www.springer.com/series/11693

Zhigang Liu • Zhiqiang Long • Xiaolong Li

# Maglev Trains

## Key Underlying Technologies

Springer

Zhigang Liu
Southwest Jiaotong University
Chengdu, China

Zhiqiang Long
National University of Defense Technology
Changsha, China

Xiaolong Li
National University of Defense Technology
Changsha, China

ISSN 2195-9862 ISSN 2195-9870 (electronic)
Springer Tracts in Mechanical Engineering
ISBN 978-3-662-45672-9 ISBN 978-3-662-45673-6 (eBook)
DOI 10.1007/978-3-662-45673-6

Library of Congress Control Number: 2015937390

Springer Heidelberg New York Dordrecht London


Printed on acid-free paper

Springer-Verlag GmbH Berlin Heidelberg is part of Springer Science+Business Media (www.springer.com)

# Preface

Maglev (derived from magnetic levitation) is a system of transportation that suspends, guides and propels vehicles, predominantly trains, using magnetic levitation from a very large number of magnets for lift and propulsion. In terms of the magnetism between magnets and magnetic components on the track, Maglev train is attracted to suspend above the track. The travel of train depends on the traction of linear motors in the vehicle. In order to ensure the safe operation of Maglev train, some key technologies must be adopted and realized, such as suspension and orientation technology, safe braking technology, position and speed detection technology, control and diagnosis technology. In the book, we present them and provide corresponding new technologies or ideas on control and diagnosis of Maglev train. We hope the contents of the book can provide some references and helps for the researcher, scholar, engineer on Rail Transit industry and Maglev technology.

The book comprises six chapters, covering about 200 pages. These chapters were written by Zhigang Liu, Zhiqiang Long and Xiaolong Li from Southwest Jiaotong University and National University of Defense Technology in China.

The book begins with an introduction of Maglev train. In Chap. 1, the characteristics and classification of Maglev train are presented. Some research and major achievements of Maglev train key technologies are discussed in detail. The technology characteristics of high-speed Maglev train and low-speed Maglev train are presented. Specially, the technology characteristics of TR08 Maglev train are discussed, and the overall design idea of TR Maglev train is outlined. In Chap. 2, the technology development and application research of Maglev control are discussed and summarized, mainly including suspension control plan and suspension control algorithms of Maglev train. Chapter 3 discusses the modeling and controller design of suspension system of Maglev train in detail. They mainly include building single point suspension model, design of control system, optimization of control parameters under the condition of output saturation, design of signal filter of suspension control system and experimental research. Chapter 4 discusses the control and diagnosis system of Maglev train, which mainly includes on-board control and diagnosis system, comprehensive assessment algorithm of on-board faults based on fuzzy comprehensive assessment, and comprehensive assessment method based on EDA (estimation of distribution algorithm). In Chap. 5, the on-board control and diagnosis networks of Maglev train are introduced. Main contents include the communication simulation of on-board diagnosis network, integrated network design of the control and diagnosis networks, diagnosis network based on ADS, control and diagnosis networks based on role automation decentralization, and on-board communication platform in Maglev train based on RTLinux. In Chap. 6, the position and speed detection technology based on loop-cable for low-speed Maglev train is introduced and discussed in detail. Main contents include position and speed detection based on the XOR pulse, position and speed detection based on sampling, and experiments and system implementation.

One important feature of the book is that some new technologies or ideas on control and diagnosis of Maglev train are proposed and discussed. The experiments presented are real-life engineering problems, as well as problems that can be helpful to apply new techniques in Maglev train.

The studies reported in this book clearly indicate an increasing interest in Maglev train control and diagnosis technologies for real-life engineering applications. These studies are expected to simulate the interest of other researchers around the world who are facing new challenges in Maglev train and key technologies.

Chengdu, China Zhigang Liu
Changsha, China Zhiqiang Long
Changsha, China Xiaolong Li

# Acknowledgments

The authors would like to thank master degree candidates Ke Huang and Song Gao and doctor degree candidates Fengshan Dou, Chunhui Dai, Song Xue, Dapeng Zhang, Ning He, Lianchun Wang and Hehong Zhang for their help in proofreading. They also wish to thank Engineers Luo Li and Qi Hongfeng for providing reference materials.

# Contents

# Chapter 1
# Maglev Train Overview

## 1.1 Introduction

Maglev (means magnetic + levitation) is a method of propulsion that uses magnetic levitation to propel vehicles with magnets rather than with wheels, axles, and bearings. With the Maglev, a vehicle is levitated a short distance away from a guideway by using magnets to create both lift and thrust. In general, Maglev trains move more smoothly and somewhat more quietly than wheeled mass transit systems. Their non-reliance on traction and friction means that acceleration and deceleration can surpass that of wheeled transports and they will be protected from the weather. At very high speeds of the conventional wheeled trains, the wear and tear from friction along with the hammer effect from wheels on rails will accelerate equipment deterioration and prevent mechanically based train systems from routinely achieving higher speeds. On the contrary, Maglev tracks have historically been found to be much more expensive to construct, but require less maintenance and have lower ongoing costs. Maglev can transport passengers and freight over long distances at speeds of hundreds of miles per hour. Maglev promises to be a major mode of transport in the twenty-first century, even more important than autos, trucks, and airplanes [1].

The term "Maglev" refers not only to the vehicles but also to the railway system, specifically designed for magnetic levitation and propulsion. All operational implementations of Maglev technology have the minimal overlap with wheeled train technology and have not been compatible with conventional rail tracks. Considering the fact that they cannot share existing infrastructure, these Maglev systems must be designed as complete transportation systems. In addition, the vacuum tube train systems might hypothetically allow Maglev trains to attain speeds in a different order of magnitude. While no such tracks have been built commercially yet, there are efforts being made to study and develop "super-Maglev" trains.



Z. Liu et al., *Maglev Trains*, Springer Tracts in Mechanical Engineering,
DOI 10.1007/978-3-662-45673-6_1

## 1.2 Characteristics and Classification of Maglev Train

There are two particularly notable types of Maglev technology [2, 3] as follows.

For electromagnetic suspension (EMS), the electronically controlled electromagnets in the train attract it to the magnetically conductive (usually steel) track.

For electrodynamic suspension (EDS), it uses superconducting electromagnets or strong permanent magnets, which create a magnetic field that induces currents in nearby metallic conductors, when there is relative movement which pushes and pulls the train toward the designed levitation position on the guideway.

Another experimental method is called as the magnetodynamic suspension (MDS), which uses the attractive magnetic force of a permanent magnet array near a steel track to lift the train and hold it in place. This method was designed, mathematically proven, peer reviewed, and patented, but is yet to be built. Other technologies such as repulsive permanent magnets and superconducting magnets have been researched recently.

Maglev trains can be classified into EMS trains and EDS trains according to their suspension mechanism. For EMS train, the onboard suspension electromagnet (or permanent magnet plus excitation control coils) that is placed under the track is energized to generate the electromagnetic field through the mutual attraction between the electromagnet and the rail. The vehicle is suspended on the rail. The suspension gap between electromagnet and electromagnetic rail is about 8–10 mm. The train is drawn by the linear motor, and the suspension gap is kept steady by controlling the excitation current of suspension electromagnet.

The main advantage of suspended Maglev systems is that they can work at all speeds and avoid the disadvantage of electrodynamic systems which only work at a minimum speed of about 30 km/h, which can eliminate the requirement for a separate low-speed suspension system and can simplify the track layout as a result. However, the dynamic instability of the system puts high demands on the tolerance control of track, which would actually offset or eliminate this advantage. In order to make a track with the required tolerances, the gap between the magnets and rail would have to be increased to the point where the magnets would be largely unreasonable [4]. In practice, this problem was addressed through the increased performance of feedback systems, which allow the system to run with the close tolerances.

The EDS train is suspended only when it reaches a certain speed. When the train is running, the moving magnetic field of onboard magnet will excite the induced current in suspension coils (always low-temperature superconducting coils or permanent magnets) installed in the line. The moving field interacts with the induced current and produces an upward force to draw the car upon the road surface by a certain height (10–15 cm generally). The train is drawn by the linear motor. Compared with EMS Maglev train, EDS train cannot suspend in stationary state or under a certain speed (about 150 km/h). Under the general speed, EDS system has a larger suspension gap, and the active control for the gap is unnecessary.

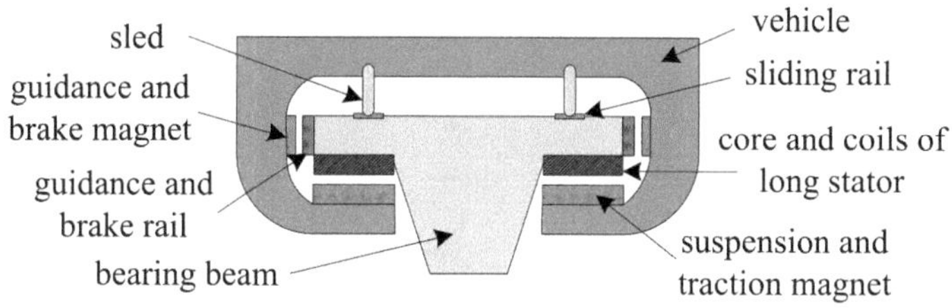


**Fig. 1.1** Principle of German TR Maglev train

The main advantage of EDS Maglev systems is that they are naturally stable minor narrowing in distance between the track and the magnets, which creates the strong forces to repel the magnets back to their original position. Meanwhile, a slight increase in distance greatly reduces the repulsive force and returns the vehicle to the right separation again. In addition, the attractive force varies in the opposite manner and provides the same adjustment effects. No feedback control is needed. At the slow speed, the resultant magnetic flux and the current which is induced in these coils is not large enough to support the weight of Maglev train. For this reason, the train must have wheels or some other form of landing gear to support the train until it reaches a speed that can sustain the levitation. Since a train may stop at any location, due to equipment problems, for instance, the entire track must be able to support both low-speed and high-speed operations. Another disadvantage is that EDS system naturally can produce a field in the track in front and to the rear of lift magnets, which can act against the magnets and produce the drag force [5].

The latest EMS Maglev train is represented by German TR08 and Japanese HSST100L. The suspension of TR08 is realized by the attraction power between the suspension electromagnet installed on vehicle suspension frame and long stator core placed on the railroad. The guidance is realized by the attraction power between the guidance magnet installed on the suspension frame and the lateral guidance rail. The moving magnetic field from the charged long stator synchronous linear motor can draw the onboard suspension electromagnet to move along the rail and realize the attraction and braking functions of Maglev train. Unlike the TR08 train, the HSST100L train is suspended and guided by the same set of electromagnets, and its traction is realized by the onboard linear induction motor, which is mounted on the rail secondarily. The principles of suspension, guidance, and traction of two types of Maglev trains are indicated in Figs. 1.1 and 1.2.

HSST Maglev train is guided by the transverse force component generated when the electromagnet moves off the inverted U shape railway center line. Since this force is relatively small, it works only when the speed is relatively low. The guidance principle is illustrated by Fig. 1.3.

The EDS Maglev train is represented by Japanese MLX train, whose structure is shown in Fig. 1.4.

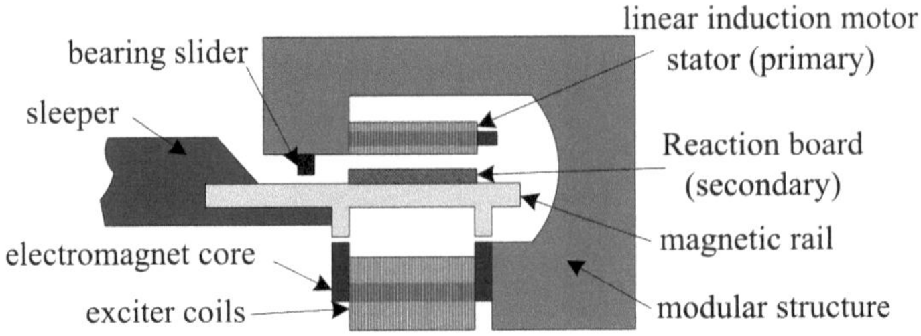


**Fig. 1.2** Principle of Japanese HSST Maglev train

**Fig. 1.3** Guidance principle of HSST

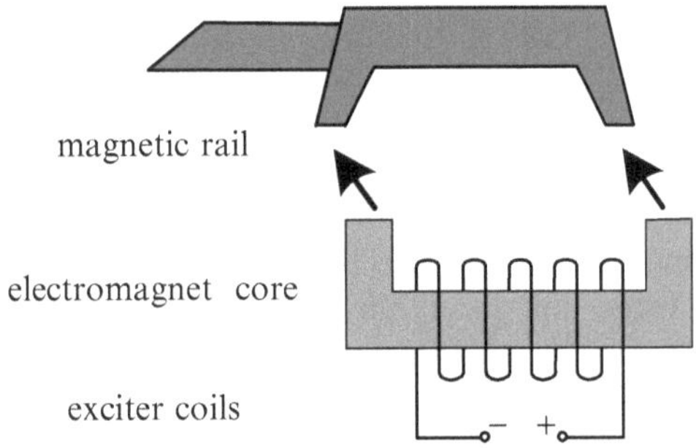


**Fig. 1.4** Transverse section of Japanese MLX high-speed Maglev train

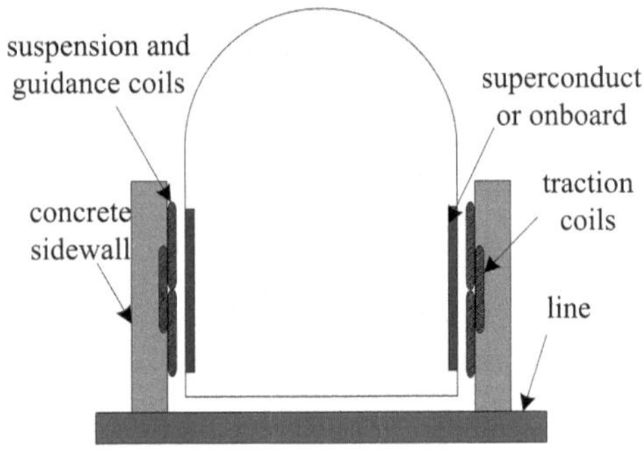


The linear motor is the only type of motor used in traction of Maglev trains, and it can be divided into two types, namely, long stator linear synchronous motor and short stator linear induced motor [6]. When the long stator synchronous motor is applied, the stator (also called the armature) is spread out along the whole line and the rotor (also called the excitation pole) is mounted on board. When the short stator linear induced motor is applied, the stator (the primary) is mounted on the board and the rotor (the secondary) is mounted on the rail. For the long stator linear synchronous motor Maglev train, the energy exchange is directly realized on the rail without contact between the car and the rail in high speed. So the long stator linear synchronous motor is more suitable for high-speed Maglev train. Both German

Transrapid normal conductor Maglev train and Japanese MLX superconductor Maglev train adopt the long stator linear synchronous motor as the traction motor.

Japanese HSST low-speed Maglev train adopts the short stator linear induced motor as the traction power. The short stator linear induced motors have the relatively simple structure and lower manufacture cost, but its efficiency and power factor are relatively lower. Due to the fact that the power supply is realized by the contact between ground devices and motive cars, the entirely contactless operation is unachievable. Therefore, the short stator linear induced motor is more suitable for low-speed (the maximum speed of about 100 km/h) Maglev trains.

German Transrapid Maglev train and Japanese HSST Maglev train can also be called as the ordinary conductor Maglev train because they adopt the charged ordinary conductor to excite magnetic field and generate suspension force and guidance force. Japanese MLX Maglev train adopts the low-temperature superconductor coil mounted on board to produce strong magnetic field, and therefore it is called as superconductor Maglev train.

For the operation speed, German Transrapid Maglev train has a maximum speed of 400–500 km/h. Japanese MLX superconductor Maglev train speed is 500–550 km/h. Japanese HSST Maglev train speed is 100 km/h. German Maglev train and Japanese superconductor Maglev train are also called as high-speed Maglev train, which is mainly used for trunk traffic between big cities. In addition, Japanese HSST Maglev train is mainly used for urban traffic, suburban traffic, and traffic connecting urban areas and airports, and it is called as low-speed Maglev train.

## 1.3 Research and Major Achievements in the World

In 1922, German engineer Hermann Kemper first proposed the concept of Maglev transportation and access to the Maglev technology patents in 1934. Due to time limitations of technology and process conditions, after 30 years, Maglev transportation had not been a significant development. Until 1969, the German company Krauss-Maffei developed the first Maglev train laboratory principle model (called TR01) in the world, the weight, length, and suspension of which are, respectively, 80 kg, 1 m, and 4 mm. The compartment floor is fixed for the suspension of four U-shaped electromagnets driven by a short stator linear motor-driven vehicle. In order to improve the speed of Maglev train, Germany gave up the short stator program and adopted the long stator synchronous motor traction.

In the early 1970s, the German government began to carry out the high-speed Maglev train transportation development plan. On the basis of TR01, Krauss-Maffei developed TR02 and TR04, but the system structure was relatively simple. In 1991, the German government passed the identification of TR07 and considered that the model train had reached the level of application. Germany later developed the more mature TR08 and TR09 Maglev train. TR08 was successfully applied in Germany and China co-construction plan, namely, the first commercial Maglev line (Shanghai Pudong Airport to Longyang Road subway station) in the world.

In 1974, Japan Airlines (JAL) purchased the company Krauss-Maffei patented TR04. As a basis, the low-speed Maglev train HSST was researched, and HSST-01 to HSST-05 five test models have been developed. In HSST-03, the bogie using embedded modular structure is first adopted. In HSST-04, the module structure is outsourcing. In HSST-05, through both front and rear anti-roll beams connecting, the two modules are integrated together to form a bogie. The bogie is independent of each other. The mechanical decoupling can be realized by using two internal modules of bogies through anti-roll beams, which can simplify the design and debug of suspension control system. The bogie structure of HSST-05 has been widely used in the subsequent design of low-speed Maglev train.

In order to solve the big city traffic problems, Korea began the development of low-speed Maglev train very early. In 1999, the low-speed Maglev train (UTM-01) was improved and UTM-02 was developed. In 2004, Korea began a study for the commercial operation for low-speed Maglev train based on UTM-02, and in March 2006, the low-speed Maglev train line engineering operation was launched. In 2007, after comparing the three major cities of Korea, Koreans eventually chose the line of Incheon International Airport as the low-speed Maglev demonstration line. At present, the first phase of construction work has been completed.

In the light of recent emergence of urban traffic congestion, the US Department of Transportation began to study the possibility of low-speed Maglev transportation, and the Maglev train technology had been included in the US Department of Transportation development plans. The US Department of Transportation supported the Maglev development project that included five class different technical contents. The current US Maglev technology is in the development phase of the prototype.

The Maglev technology has been researched for more than half a century. So far, many countries are still conducting the active research on Maglev technology, which mainly includes low-speed Maglev trains used in the urban rail transport, high-speed Maglev trains used in the intercity rail transport, Maglev momentum flywheels, Maglev reaction flywheels, Maglev energy storage flywheels, Maglev gyro torques used in satellites and space stations, and high-speed rotating machinery supported by different kinds of magnetic bearing such as various types of pumps, centrifugal machinery, grinders, etc. Besides, some high-precision and high-sensitivity instruments such as Maglev gyros and Maglev gravimeters also adopt Maglev technology [7–12].

It is known that the Maglev system can be divided into the electromagnetic suspension (EMS) system and electrodynamic suspension (EDS) system based on the different generating principles of Maglev forces. The EMS system uses attraction force between charged solenoid and rail to make trains suspend. The attention of representative studies in countries like Germany, Japan, China, Korea, and Sweden is focused on the development of Maglev trains and magnetic bearings [12]. The German high-speed Maglev train TR08 has already been in commercial operation line from Shanghai Longyang Road to Pudong airport [13]. China began to research the Maglev transport technology in the 1980s. At present, the self-developed low-speed Maglev train operates well in Tangshan test line. The EMS-type Maglev train adopts the electromagnetic interactive force between onboard electromagnets and rails to realize the no-contact support and the guidance of trains. The steady suspension of vehicles needs multiple specialized suspension controllers, which have complex structure [14–16].

For the EDS systems, the suspension derives from the repulsive force generated by the interaction of source magnetic field and induced current, which is produced by the relative motion between the permanent magnet/superconducting magnet and the rail. The magnetic resistance force is generated in the same way. The EDS Maglev system technology is represented by the low-temperature superconducting magnet high-speed Maglev train of Japan and permanent magnet low-speed Maglev train of the USA. The EDS systems technology can be divided into two types that are superconducting magnet type and permanent magnet type according to the source magnetic field [17]. At present, the research of the EDS technology is focused on the superconducting type because the permanent magnet cannot provide enough field strength.

With the advent of permanent magnetic material with high permanence intensity and the continuous application of the Halbach structure permanent magnet array, the permanent magnet can provide enough suspension force and has been gradually introduced into the EDS system to replace the superconducting magnet. The permanent magnet-type technology does not need the cooling devices and can achieve a high ratio of suspension force to resistance force at certain speeds. However, for the permanent magnet-type system, the active control must be introduced to keep the system stable because the permanent magnet is characterized by critical stability and is easy to oscillate when the disturbances occur [18–20].

## 1.4 Research and Major Achievements in China

The EMS-type Maglev systems have a wide application area, among which four applications (low-speed trains, high-speed trains, magnetic bearings, and Maglev flywheels) have adopted the Maglev system control. The low-speed Maglev train has a maximum speed of 100–130 km/h and is mainly used in the city transport. The cost is fairly equal to the light rail, and its most important advantage is clean and green. The high-speed Maglev train has a maximum speed of 400–500 km/h and is suitable for intercity transport.

Although there is a large difference between low-speed and high-speed Maglev trains in traction, guidance, power supply, and operation control, their suspension principle is similar, for that both of them use the active electromagnetic suspension principle and both suspension gaps are around 10 mm. Both of them use the attraction force between the controlled suspension electromagnet installed under vehicles and the electromagnetic reactive rail installed on the base of guide rails, to make the vehicle suspend. Therefore, there are many similarities in the Maglev control systems for low-speed and high-speed Maglev trains.

Both the magnetic bearing and Maglev flywheel use the magnetic force to make rotators float in the space and have the advantages of no friction, no need of lubrication, strong resistance against environment, lower heat, high precision of gyration, and high rotation speed. In order to satisfy the high-precise satellite control, the attitude control of large space vehicle (e.g., space station), and the need of high-speed ground rotating machinery, many countries carried out researches in magnetic bearings and Maglev flywheels supported by magnetic bearings. Of course, the magnetic bearing also has a complex structure and the controller of the general 5-axis active suspension control magnetic bearing has problems of decoupling and fault-tolerant control, etc.

The low-speed Maglev train generally uses the noncontact operation mode. Compared with the normal wheel–rail trains running within the city, Maglev trains have advantages of lower noises, smaller vibrations, looser track laying conditions, lower building costs, and easier complement and maintenance. In addition, Maglev trains have stronger climbing ability and smaller turning radius since the traction force of Maglev trains is not affected by the adhesion coefficient between wheel and rail.

Since the 1980s, some colleges and research institutes in China, such as Southwest Jiaotong University, National University of Defense Technology, Institute of Electrical Engineering of Chinese Academy of Sciences, and Chinese Academy of Railway Sciences, etc., have carried out some basic researches on Maglev trains.

In July 2005, an engineering prototype vehicle was manufactured in Tangshan Railway Vehicle Factory. A practical Maglev train marshaling of two cars was developed and carried out an operation test on a 1,547-m-long test line in Tangshan in May 2009, with a maximum speed of 105 km/h. National University of Defense Technology and other units started to develop the hybrid magnet Maglev train and successfully developed one in December 2012, which has already been put into the actual operation test.

Since December 1998, Southwest Jiaotong University has built a 425-m-long railway test line in Qingchengshan Chengdu and developed a low-speed electromagnet Maglev train in the Changchun Railway Vehicle Factory in October 2001. The train was tested on Qingchengshan test line in Aril 2006.

In order to promote the application of low Maglev trains in Shanghai urban transport, Shanghai Research Center of Maglev Transport Technology united with other related domestic units built a 1,500-m-long low Maglev test line and developed a low Maglev train.

Otherwise, the CSR Group (China Southern Railway Corporation Limited) also began to study the technologies of low-speed Maglev trains. They built a test line in the Zhuzhou Railway Vehicle Factory and developed a low-speed Maglev test vehicle, which are now in the stage of development [21–24].

In October 2000, a high-speed Maglev transport line from Longyang Road, Pudong, Shanghai, to the Pudong International Airport was built, and this line started to operate in September 2003. In this context, in order to master and improve the key technology of high-speed Maglev trains, with the support of the Ministry of Science, organized by National Maglev Engineering Center, China carried out the researches of high-speed Maglev transport and developed the technologies of permanent-electromagnetic hybrid magnet for the high-speed Maglev train.

All of the above shows that China has basically mastered the key technologies of Maglev train and has developed preliminary prototypes, engineering prototypes, and practical vehicles successively. There are application requirements and technology basis to develop the Maglev transport system in China.

## 1.5 Development and Discussion of Maglev Key Technologies

As a new transport tool, Maglev transport still has many problems involved in many subjects to solve [25–29]. Considering only the Maglev control system, the problems include vehicle–track coupling dynamic modeling [30, 31], vibration control [32–34], suspension control [35, 36], fault diagnosis and tolerance control [37–41], hybrid magnet study, etc. [42, 43].

### *1.5.1 Low-Speed Maglev Trains*

#### 1.5.1.1 Coupling Model Establishment and Vibration Analysis

The vehicle–track coupling vibration problem of EMS Maglev train is a problem of the current process of commercialization of the Maglev system to be solved. In [29], the self-excited vibration occurrence condition of the coupling system is determined based on Nyquist theory in frequency domain. The nonlinear model rail coupling system is studied in [33]. Maglev train vehicle–track coupling system stability is investigated when there is a delay of signal feedback path through adopting the center manifold reduced-order and multi-scale method in [44]. Some new methods and conclusions on Maglev track modeling and vehicle–track coupling self-excited vibration mechanism are proposed in [45]. The coupled dynamics of Maglev train vehicle–track is investigated in detail and was compared with the wheel–rail system in [36]. In addition, the track coupling dynamics problem of Maglev train vehicle–track is also studied in [46].

In order to overcome the self-excited vibration occurrence of Maglev train vehicle–track coupling, there are two common methods, including increasing the stiffness of track and adjusting the parameters of suspension controller. However, they have certain limitations. How to find a methodology, which cannot only reduce the cost of the track but also to avoid the vehicle–track coupling self-excited vibration, is the hot spot area of Maglev technology.

#### 1.5.1.2 Suspension Control Algorithms of Maglev Train

The suspension control is always the technical difficulty of EMS Maglev train. The suspension control algorithm should meet the following requirements:

- Good adaptability for the uncertainty suspension system model
- Good robustness for the interference of the external system
- Good inhibiting ability for the vehicle–track coupling vibration

In [47], the suspension control system is decomposed into sub-current loop control system and suspension of two serial and decoupled subsystems to consider a fast current loop design approach, which can provide the foundation for Maglev control engineering design.

In order to overcome the trouble from the linearized design, the gain adjustment control method is adopted, which can linearize multiple work nodes and obtain multiple linear models within the work scope of the suspension system. In order to overcome the problems of control algorithm design, such as parameter perturbation, external interference, and vibration aspects of vehicle–track coupling, more and more modern control methods have been applied to the design of the suspension control algorithm. These methods include optimal control, fuzzy control, neural network control, adaptive control, robust control, and sliding mode variable structure control, etc.

These methods have strong inhibitory effect for suspending load changes, external disturbance, control object parameter drift, and other factors, but have high complexity, and there are some difficulties in the application of some methods.

#### 1.5.1.3 Fault Diagnosis and Tolerant Control of Suspension Control System

In order to achieve the vehicle suspension, a five-bogie low-speed Maglev train is equipped with 20 sets of general suspension controllers, 20 clearance and integrated acceleration sensors, and 40 electromagnet groups. Whether it is for low-speed, high-speed Maglev train or electric and electromagnetic hybrid suspension train, once some controller, sensor, or electromagnet (actuator) of the suspension system fails, the corresponding suspension points will be suspended instability or failure [48].

For the comprehensive evaluation of suspension system failure, since the model parameters of traditional fuzzy comprehensive evaluation method to assess the fault level problem are difficult to be determined, a fuzzy comprehensive assessment method based on the distribution estimation algorithm is proposed in [49]. In this method, the model parameters are automatically optimized by using evolutionary computing parameters, which can improve the assessment accuracy and retain the advantages of traditional fuzzy comprehensive evaluation model.

In the device-level fault diagnosis and fault-tolerant control of the vehicle suspension system, the fault diagnosis methods based on model and signal are respectively adopted. In [50, 51], aiming at the uncertain nonlinear suspension control system, a feedback control and fault diagnosis-integrated design plan is proposed. The method can make the system robust to the uncertain model and have a strong track of sensor failure at the same time. In addition, the sensor failure of Maglev network control system is analyzed by simulation. In [52], the tracking differentiator for Maglev train accelerometer fault diagnosis strategy is proposed, and the nonlinear discrete tracking differentiator is adopted to extract the higher reliability gap differential signals. The signals are a benchmark to determine whether the acceleration integral signal is abnormal.

#### 1.5.1.4 Permanent and Electromagnetic Magnet Hybrid Suspension Control Technology

On the basis of the EMS study, new Maglev technology is researched and actively explored, which has made great progress in permanent and electromagnetic magnet suspension system, called “PEMS” system. The technology is different from that in the electromagnet embedded permanent magnets, and the permanent magnets can share most of the vehicle weight, and the electromagnetic coil current can play only a dynamic regulatory role. Therefore, PEMS technologies can significantly reduce the suspension current situation to improve the heating situation of magnets.

In [53], based on the analysis of magnetic levitation mechanism, the idea of permanent magnet Maglev train is proposed, and the suspension controller with zero power is designed.

It is reported that the PEMS vehicle suspension system has achieved a full range of stable operation in the 1.5 km Tangshan test line [54, 55]. The suspension system in power dissipation, electromagnet temperature rise, temperature controller, and other aspects have greater advantages.

### *1.5.2 High-Speed Maglev Trains*

#### 1.5.2.1 Lap Structure Suspension Fault-Tolerant Control Problem

In order to avoid the individual point suspension failure that leads to the train unable to achieve full vehicle suspension problems during the high-speed operation, a suspension lap structure is adopted for high-speed Maglev train. Dual electromagnet model based on the elastic coupling is adopted to represent the overlapping structural model. Based on nonlinear control theory, the nonlinear decoupling controller is designed for suspension lap structure. In [50], a more accurate mathematical model structure suspension lap is established. By analyzing high-speed Maglev train control objectives of overlap structure, the lap controller design idea of the suspension structure as a whole is proposed. The state feedback is adopted in the controller design with linear structure, which is applied in the suspension control of high-speed Maglev train overlapping structure.

For the fault tolerance of suspension control system, the passive fault-tolerant controller of suspension system of Maglev train based on overlapping structures is designed for the open-loop unstable linear system. In addition, aiming at the problem that the clearance measurement unit cannot have effectively redundant and prone failure, the state observer method is proposed to solve the problem of redundant clearance measurement unit. Meanwhile, due to the fact that a suspension control unit cannot work for the suspension overlapping mechanism, the suspension control strategy is given.

#### 1.5.2.2 Active Guide Control Problem

The guide force of high-speed Maglev train active guide system could vary over a wide range, which is influenced by the tracks and external interference. Therefore, for the guidance system in the normal course of their work, their work point current will show large-scale changes. In [50], more accurate guide system model is built, and the practical guide controller is designed based on the robust control theory and adaptive control theory. For smaller range of operating point current changes, the robust control theory is adopted to design the guide robust controller. For the wide range of operating point current changes, the fuzzy control theory is adopted to design the fuzzy gain scheduling mechanism, which can overcome mutation and oscillation problems in the switching process.

#### 1.5.2.3 Gap Measurement Problem

The suspension guide gap detection of EMS high-speed Maglev train is realized by one kind of inductive displacement sensor. The slot effect is a special problem based on the suspension clearance measurement of EMS high-speed Maglev train. The inductive sensor based on FPGA is adopted to detect the suspension gap, and an approach, which the detection coil width is equal to one cycle, is proposed to weaken the slot effect. The same gap detection method is adopted to detect the suspend gap, and the magnetic induction power of detection coils is analyzed, which is produced in space. The method through changing the magnetic field spatial distribution of the detection coil is proposed to weaken the slot effect. The changes caused by magnetic field distribution for the detected objects are analyzed. In [49], based on the magnetic mirror method for detection coils, the equivalent inductance coil is analyzed, and the slot shape function is introduced. Then, the relationship between coil geometry parameters and slot effect is studied. But the method is more complex and more difficult to the practical analysis and application.

#### 1.5.2.4 Speed and Position Problem

The EMS high-speed Maglev train is actually one kind of linear synchronous traction systems. In the low-speed sections, it is necessary to obtain accurate secondary polar phase angle signal with positioning equipment to achieve steady traction control. Meanwhile, some problems such as the suspension gap fluctuations, long stator rail joints, which may affect the normal operation of the special problems, should be considered. The detection of vehicle relative position depends on the sensitive count between slot structure and relative position sensors.

In [52, 53], the count method of slot cycles with inductive sensors for long stator is adopted, and the slot cycle subdivision method with speed linear interpolation is used to solve the problem positioning accuracy. In [54], using the detection coil equivalent inductance characteristics similar to a sinusoidal variation, the inverse trigonometric calculation method is adopted to achieve high-precision positioning. In [55], based on the fast adaptive filtering forecast, the switching method of two-way detection signals and fast differentiator method are proposed, which can eliminate the signal distortion caused by rail joints and meet the requirements of the relative position detection.

## 1.6 Technology Characteristics of High-Speed Maglev Train

### *1.6.1 Technology Characteristics*

As a rail transport system, German Transrapid Maglev train system has the following technical characteristics [3, 14]:

1. The suspension and guidance functions are realized through controllable electromagnets without the contact and wear between vehicles and rails.
2. The no-contact traction and braking are realized by long stator synchronous linear motors. The control and transmission of traction power are conducted by ground devices. The stator of long stator linear motor is mounted on the rail.
3. The electrical energy required by the train in operation is supplied without the contact.
4. Maximum application speed is 400–500 km/h.
5. Higher acceleration capturing ability is better than conventional railway system.
6. More flexibility in design line parameters; smaller tuning radius and higher climbing ability than conventional railway under the same speed.
7. The vehicle embraces the track, and therefore there is a lower derailment risk than conventional railway.
8. The noise of Maglev train is lower than cars or conventional trains under the same speed.

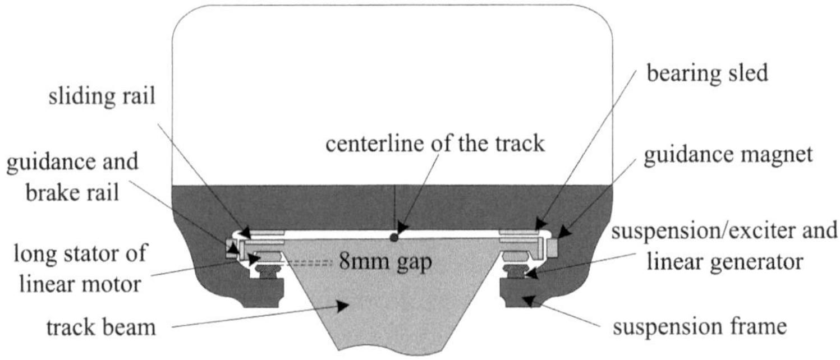


**Fig. 1.5** Transverse section of Maglev train and track beam

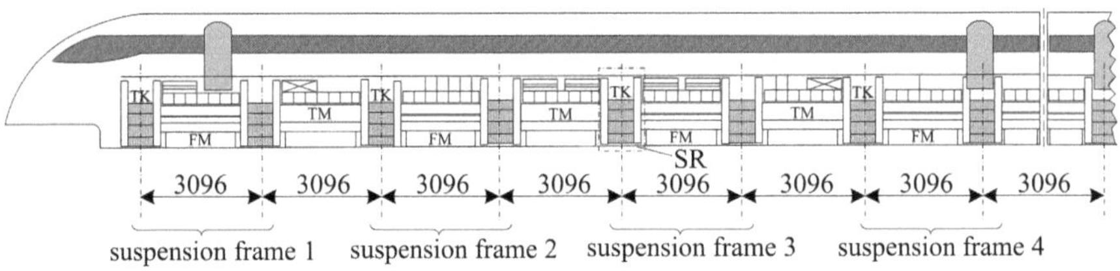


**Fig. 1.6** Positions of main functioning parts of TR08

## 1.6.2 Suspension and Guidance System

The individually controlled electromagnet mounted on Maglev train can draw the electromagnetic rail. The stator core mounted on the base of the line downward can make the vehicle overcome the gravity and keep suspended. The guidance magnet interacts with the guide rail to generate a transverse force, which can make the vehicle keep consistent with the central line of rail, further to realize the vehicle guidance. The suspension gap is kept 8–10 mm by highly reliable suspension control system. The arrangements in transverse and longitudinal section of TR08 Maglev train are respectively shown in Figs. 1.5 and 1.6.

## 1.6.3 Long Stator Synchronous Linear Motor Traction

The long stator synchronous linear motor is used for the traction and braking in Transrapid high-speed Maglev system. By charging the linear motor stator laid along the line with three-phase alternative current, a magnetic field moving along the line is generated. The excitation components in Maglev trains are drawn by this field to make the vehicle move forward synchronously with the moving magnetic field. The traction and braking force are adjusted by controlling the amplitude and

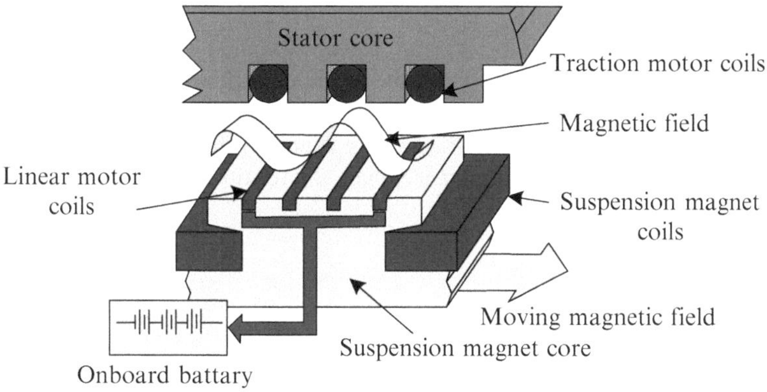


**Fig. 1.7** Suspension electromagnets and the linear motor

frequency of stator coil current, which can realize the continuous adjustment of speed from zero to rated operation speed. The traction principle is shown in Fig. 1.7.

Different from conventional transport system, the traction motor stator of TR08 Maglev high-speed system is mounted in the line instead of on board. The motor power can be adjusted based on the actual needs in operation.

### *1.6.4 Braking*

The braking of the Transrapid Maglev train is realized by long stator synchronous linear motor in normal operation. When the train is braking, the linear motor switches from traction state to generating power state and the kinetic energy of the vehicle is transformed into electrical energy to feed back to the power grid by resistance. Simultaneously, the vehicle is decelerated with the electromagnetic resistance force until the vehicle stops.

When there is emergency brake due to system fault or other security problems, the braking is realized by the eddy current brake mounted in each Maglev car. The eddy current brake is charged by linear motors onboard (when in high speed) and batteries, with no relying on the traction or power supply system on the ground. When the eddy current brake is in operation, the security system calculates a braking process curve according to the distance between the transient location and the nearby parking lot. The braking force can be controlled by the input current of eddy current brake according to the curve. The current in the braking coils can excite a magnetic field which interacts with the guide rail to generate a resistance force for the vehicle.

As the speed of vehicle gets lower, the attractive force between the eddy brake and the guide rail gets larger and finally large enough to make the surface of eddy brake contacting with the guide rail and generating a lateral resistance force against the operation direction. The force that is combined with the electromagnetic resistance together would decelerate the vehicle. When the speed is less than 10 km/h, the eddy current brake is shut down and the vehicle will descend on the sliding rail and park in a scheduled area with the onboard sled that bears the whole vehicle. It will slide for a short distance under the friction between the sled and the slide rail.

### *1.6.5 Vehicle–Rail Relationship*

There are three pairs of important functioning interactive surfaces, namely, the pairs between the stator core surface on the bottom side of two rail wings and the surface of suspension magnets on board, the pairs between the surface of the guide rail in two lateral sides of the line and the surface of guide magnet (or eddy current brake) on board, and the pairs between the surface of the sliding rail on the top of two rail wings and the supporting sled on board (Fig. 1.5).

The spatial location and relation of three functioning surfaces are decided by the line selection, structure design, and installation, which will introduce the inevitable irregularity and geometrical deviation. The temporal and permanent deformation would appear under environmental loads (temperature difference, wind, etc.) and operation load. For all of above problems, the vehicle should be able to adjust these deviations and irregularities to pass the curve without contacting to the rail. With the vertical and transverse vehicle–rail impact that can efficiently be insulated or decayed, an environment of traveling comfort should be ensured.

In order to adapt to the terrain, the curved section and ultrahigh of line are introduced in this section. In the transition curve section, the function surface may twist and the electromagnet on board and bogie may also twist. The magnet on board is mechanically divided into sections of about 3-m length. These sections allow a small angle between the adjacent sections and sections on the opposite side and have a resilient connection with the suspension frame to make the vehicle follow the line curve without contact. In other words, the boundary parameter of line selection and geometric deviation of rail function surface are decided by the minimum length (about 3 m) of rigid bodies on board, tolerance of suspension gap, vehicle guidance system, mechanical suspension system, and comfort requirement.

Since Maglev train adopts the electromagnetic force to guide and suspend, it generally has a smaller turning radius than the wheel–rail train in the same speed. However, the turning radius of Maglev train is larger than the wheel–rail train in low speed (less than 50 km/h) because of the vehicle bogie restriction.

Considering the sufficient riding comfort and noise suppression, the improvement of vehicle capability, such as reducing the requirements of the track, has been an important controversial issue and a comprehensive research issue of technology, system, environment, and economy.

### 1.6.6 Power Supply for Onboard Devices

In normal operation, the power consumption of all onboard devices of Maglev train, such as suspension, guidance, air conditioning, illumination, communication devices, etc., is about 160 kW. When the operation speed is less than 100 km/h, Maglev train is powered though the contacts between the DC power rail and the current collector on board. When the speed is higher than 100 km/h, the power demand of onboard devices can be satisfied by the linear generator mounted on suspension magnets. In this case, Maglev train and the track are completely noncontact. When the train operates normally, the battery on board is full and in charging status. When the system breaks down and the eddy current brake needs to work, the battery will provide the energy to brake the train.

### 1.6.7 Positioning and Mobile Communication

When the train is running, it is necessary to make the control system sense its exact location and speed. Hence, a set of positioning system should be attached in the train. The function of positioning system is realized by devices installed in the train and track.

A coding board based on the absolute position is installed along the line for each certain distance. When the train passes by these positions, the reader mounted on board will read the coding board to obtain the absolute position of the train. The absolute positioning device is installed redundantly. Besides, a set of electromagnetic devices which count the long stator alveolar is put on the vehicle to confirm the vehicle position more accurately (the obtained position is the position relative to the nearest coding board), with an accuracy of the width of a tooth or a groove, namely, 43 mm. The positioning information can be used to calculate the vehicle speed. The suspension Maglev pole position relative to the long stator synchronous linear motor stator magnetic field can be derived from the code value. The quantity can be used in the control of low-speed linear motor (synchronous motor rotor flux-oriented control).

All positioning and speed data are firstly obtained on board, and then they are transmitted to the ground operation control and traction system for conducting the operation of trains. Therefore, the mobile communication between the ground and the moving vehicle is needed. Both German TVE Maglev test line and Shanghai line adopt the 38G radio communication to transmit data for the vehicle control, security, and synchronous traction. The passenger information and communication are transmitted through another radio communication channel.

### *1.6.8 Overall Design Ideas of TR High-Speed Maglev Train*

#### 1.6.8.1 Redundant Design for Safety and Reliability

The support, guidance, traction, and braking function are realized by vehicle–rail contact in conventional railway. The power supply depends on the catenary–pantograph contact. For Maglev train, electromagnetic force and electromagnetic induction can realize these basis functions and avoid the contact between the high-speed vehicle and the rail. Compared with the conventional train, Maglev train greatly reduces the number of mechanical wear components. However, the number of electric and electronic devices will grow. For example, TR08 train has about 330 pieces of the minimum-sized substitutable electronic devices and electromagnet modules which are 1/3 of vehicle weight. Since the time between failures of a single electric or electronic component is finite, an essential design rule, which ensures normal operation when a single electric or electronic part fails, should be established to ensure the operation safety and reliability in Maglev train development progress.

The parts for essential functions are assembled in a modular, independent, and redundant way. When any single part fails, the train can still keep running by schedule as normal. Only combined fault with multi-parts could interrupt the normal operation of the train. The estimation of availability rate of Maglev train is more than 99.99 %, of which the number depends on the train number and the plan of operation and maintenance.

To illustrate the principle of redundancy, a few specific examples are listed in the following:

Example 1: Safety and reliability of power grid are achieved based on principles of dispersion and independence.

The TR08 train adopts DC 440 V as the onboard power supply. The electric and electronic devices of each car are fed by four independent grids that are insulated from each other. The power consumption and spatial location of devices on each grid are balanced. Each grid is powered by eight booster choppers (two grids in the end car only have seven booster choppers). When three of the booster choppers fail, the grid could still work. Even if the grid fails and the relevant suspension magnets lose power, Maglev train can still keep suspended until the train enters the nearest parking area, which means that failure of one single or even more than one power supply devices will not result in the operation interruption of Maglev train and the failed components could be replaced in the time quantum of the operation or after a day's operation.

Example 2: Safety of onboard eddy current brake.

Each car has a pair of eddy current brakes which totally consist of 24 magnet poles. They are divided into eight groups and each group is composed of three series poles. The eight groups of poles are separately connected to four grids. Even if one grid fails or four of the eight brake circuits fail, the brake can still work safely.

Example 3: Safety and reliability of suspension system.

Each running gear of Maglev vehicle consists of four suspension frames, each of which has 4 bars. Namely, each vehicle has 16 suspension bars. The suspension electromagnet of about 3-m length and the guidance electromagnet are connected to adjacent suspension bars. Two independent suspension circuits provide the levitation force for each bar. Each suspension circuit consists of a part of an electromagnet, two suspension gap sensors, and one suspension control unit.

When one control unit or the relevant two gap sensors fails, the corresponding air spring will be emptied and the load of the suspension frame will be reduced. In this case, Maglev train could still keep running, and the failed components could be fixed in the time quantum of the operation or after a day's operation. When the two suspension control units or the four gap sensors (small probability) for the same suspension frame fail, the relevant air spring will be emptied and the vehicle will lose support. Then the train will land on the rail and slide by sleds to the nearest station and be drawn to the maintenance center.

#### 1.6.8.2 Online Diagnosis System

Each Maglev car has about 330 electric components and almost all of them have online diagnosis function. The maintenance is based on the automatic fault diagnosis of components. The train operation status can be obtained at all times from the onboard diagnosis computer and the ground maintenance center. When operation failure appears, the operation and maintenance personnel can receive the fault report quickly, by which they are able to let the train continue to run or arrange the train to exit when the operation density is low.

#### 1.6.8.3 Equipment Layout and Installation

Most of the electric components are installed in the sandwich structure of vehicle body like drawers. Thus, a failed component can be simply replaced in 15 min by opening the apron, pulling out the failed component, pushing into a spare component, and locking the apron. When the replacement is finished, the diagnosis system will immediately display the fault status. Through the special equipment layout and installation manner, the maintainability of the vehicle is improved and the maintenance work is greatly reduced.

#### 1.6.8.4 Electromagnetic Compatibility, Sealing, and Vibration Suppression of Electronic Devices

In order to avoid the electromagnetic interference and achieve electromagnetic compatibility requirements, most electronic devices are sealed in the cast drawer-type chassis, and most cable connectors are sealed in the cast connecting box. These boxes are also designed to prevent moisture and ash which can cause insulation failure. For the components that are sensitive to vibration, such as the suspension gap sensor, the rubber cushion isolation in connecting parts is needed.

#### 1.6.8.5 Car Body Lightening Design

The electromagnetic attractive force supports the weight of car body. Therefore, the weight greatly affects the power consumption of Maglev train. The structure size of long stator and the temperature rise of suspension magnet will also limit the vehicle weight. The vehicle weight consists of self-weight and payload. In order to save energy and increase payload, the self-weight should be decreased. Therefore, lightening the car body has been an important design rule [6]. The total weight could be reduced only by strictly controlling the weight of every single part including the mechanical, electrical, and electronic part.

## 1.7 Technology Characteristics of Low-Speed Maglev Train

Take HSST low-speed Maglev train as an example for low-speed Maglev trains and its principle structure is shown in Fig. 1.8. Its suspension and guidance functions are realized with a U-shaped electromagnet and the inverted U-shaped track, and its traction adopts the linear induction motor (LIM).

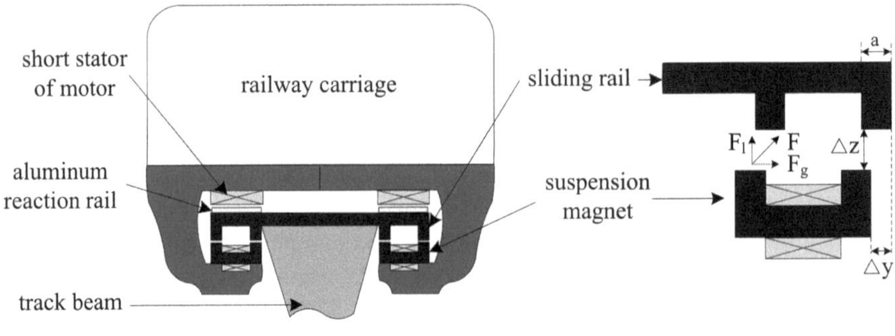


**Fig. 1.8** Suspension electromagnets and the linear motor

### 1.7.1 Suspension and Guidance Function Realization of Low-Speed Maglev Train

For HSST low-speed Maglev train, U-shaped magnet is installed on both sides of the bottom of the vehicle suspension, and the inverted U-shaped magnetically rail is installed on the track. After the magnet is energized to attract the track to achieve a suspension of the vehicle, the current adjustment in the suspension magnet can make the train stably suspended in a certain clearance. In the suspended state, if the suspension produces the laterally offset relative to orbit magnet, the electromagnetic suction will laterally generate the components under the action of inverted U-shaped orbit, which can provide the guidance for the train.

In Fig. 1.8, if the U-shaped magnet produces the laterally offset $\Delta y$ and vertical clearance $\Delta z$, relative to the inverted U-shaped magnetic rail, the electromagnetic force $F$ between the tracks and magnets will provide the guidance force along the lateral component $F_g$ and provide the suspension force along the vertical component $F_l$. They can be computed with the formulas below:

$$\begin{aligned} F_l &= K\frac{i^2}{\Delta z^2}\left[1 - \frac{\pi|\Delta y|^2}{a(\pi|\Delta y|+4\Delta z)}\right] \\ F_g &= -K\frac{i^2}{\Delta z^2}\left[\frac{\pi|\Delta y|\Delta z}{a(\pi|\Delta y|+4\Delta z)}\right]\cdot \mathrm{sign}\,(\Delta y) \end{aligned} \tag{1.1}$$

where $K$ are the parameters determined by the electromagnet structure, $i$ is the current in electromagnet, and $a$ is the pole width and sign() is sign function.

Based on Formula (1.1), the guidance force $F_g$ has the opposite direction to the lateral offset $\Delta y$ and $\partial\left|F_g\right|/\partial\left|\Delta y\right| > 0$. Namely, the guidance force provides the lateral restoring force, and the force will increase with the increase of the lateral offset. Since the system is the lateral stable, it is unnecessary to equip with the guidance magnetic and controllers.

Usually, the ratio of suspension force and guidance force is adopted to measure the system guidance ability as follows:

$$\left|\frac{F_g}{F_l}\right| = \left|\frac{\pi\,|\Delta y|\,\Delta z}{a\,(\pi\,|\Delta y| + 4\Delta z) - \pi|\Delta y|^2}\right| \tag{1.2}$$

### 1.7.2 Traction Function Realization of HSST Maglev Train

The short stator LIM is adopted in HSST Maglev train traction system. The primary coils of the motor are mounted on a vehicle, and the secondary coils use aluminum rails installed in the track surface. Three-phase alternating current is applied to the stator coils to generate a traveling wave magnetic field that can induce the eddy currents. The interaction between traveling magnetic field and the eddy currents drives the train. And considering that the traction motor is located on the vehicle, some devices, such as high-voltage electrical systems, filter reactors, traction inverters, linear motor-vehicle traction, and braking resistors, should be configured.

Compared with the same capacity rotary motors, LIM efficiency is much lower. The efficiency of the rotary electric machine is generally 0.80–0.85, and the efficiency is only about 0.7 LIM. There are two main reasons. One is a large gap motors, and rotary motors generally have only 0.25–0.8-mm gap. HSST traction motor air gap, due to the constraints of the suspension gap of about 15 mm, is 10 times as much as the rotation of the motor, which results in the low utilization of air gap magnetic field. In the actual system, the security requirement for the train running is about 8-mm gap between the vehicle and suspension rail, which can ensure the motor not to touch the rail. In addition, considering the installing track error and motor installation error, the traction motor air gap is generally about 15 mm. In fact, the traction motor air gap is greater than this value, and there is almost no decrease in the space. The other is both ends of the primary core motor breaking and the stator coil is not continuous. Hence, the motor efficiency is reduced largely. Besides, the LIM power factor through the inverters in Maglev system is low and the value is about 0.6–0.7. In summary, the energy efficiency can only reach 50 % capacity, and the energy utilization rate is low.

The motor core and the primary winding of limited length will result in the presence of the input and output ends in the movement direction, which can produce the uneven distribution of the magnetic field or magnetic field offset phenomenon, known as end effect including the longitudinal end effect and the transverse end effect. The longitudinal end effect includes static longitudinal end effect and the dynamic longitudinal end effect. The core breaking and the discontinuous primary windings lead to the formation of static longitudinal end effect. Breaking the core would make each phase winding impedance unbalance. Even if there is symmetrical three-phase voltage in the winding, the winding will generate the asymmetric current, which makes the air gap magnetic field not only contain a forward traveling wave magnetic field but also contain an alternating pulse magnetic field. The pulse magnetic field will generate the additional losses and show the resistance performance in the motor road course. When there is a relative motion between the motor primary and secondary, at the front-end position of the primary forward direction, the corresponding secondary region will generate eddy currents to hamper the enhancement of magnetic field strength in the region. The corresponding secondary region at the primary rear end region will generate eddy currents to hamper the weakening of magnetic field strength in the region. For the motor performance, they mean the additional losses and additional force, known as the dynamic longitudinal end effect. Since LIM primary and secondary width is limited, the secondary current and secondary reaction plates in the limited width of the air gap magnetic field will have an impact, called the lateral side end effect, which makes the motor air gap magnetic field distort and its characteristic become poor.

The end effect of the motor will bring additional losses and additional force, which shows the resistance force in the motor road course. At a certain speed, the resistance produced by the end effect will offset the motor driving force and make the train lose the accelerating ability.

### *1.7.3 Bogie Structure of HSST Maglev Train*

The bogie structure of HSST low-speed Maglev train is used to mount the suspension and guide electromagnets and traction motors. The structure is the smallest structural unit with suspension, guidance, and traction functions. The suspension magnets are mounted on a box beam composed of magnet modules, and each module contains four magnet poles line package, a group of two, corresponding to a set of independent suspension control system.

Front and rear sides of the module by two anti-roll beams are joined together to form a bogie. The anti-roll function bogie beam provides mechanical decoupling between the four suspension points. The minimum bend radius that Maglev train can pass is limited by the length of the truck and the body lateral offset. HSST bogie has small size and large lateral space and allows greater lateral offset, which can achieve a small turning radius. The bogie length of HSST-200-type Maglev train is 3.4 m, and the corresponding minimum turning radius is 100 m. The bogie length of HSST-100-type Maglev train is 2.5 m, and the corresponding minimum turning radius is 25 m.

### *1.7.4 Technology Characteristics*

Compared with the normal wheel–rail train, low-speed Maglev system uses a noncontact operation mode, which has the following characteristics: low noise, vibration, relaxed track laying conditions, low construction cost, easy implementation, easy maintenance, etc. In addition, because the low-speed Maglev train is not affected by the wheel–rail adhesion coefficient, it has better climbing ability, small turning radius, and a unique advantage in a variety of transportation modes.

1. Small vibration, low noise, better environment protection, and comfortable ride

Since low-speed Maglev train uses electromagnetic gravitational suspended in the orbit above the vehicle, there are no noise or vibration generated by the contact between wheel and rail, and the noise of the vehicle running is very low. It is measured at the peak noise at a distance of 10-m tracks for 64 dB, and a peak corresponding conventional wheel–rail noise is 92–94 dB. There is no wheel wear, iron or rubber dust and other air pollutants, which will be helpful to the environment.

2. Small turning radius, climbing ability, and good line adaptability

The riding and braking of low-speed Maglev train do not longer depend on the adhesion between wheel and rail, instead of generating electromagnetic linear motor traction. So the vehicle has good acceleration and deceleration performance and climbing ability. Its climbing ability is up to 70‰, much higher than the 30‰ in traditional wheel–rail system.

The minimum turning radius for the wheel–rail system in the ordinary line is 300 m. The turning radius of low-speed Maglev train may be small, and the minimum turning radius in the ordinary line is about 75 m. Even in a large road intersection in the city, the turning can be completed easily. Therefore, the low-speed Maglev train rail lines are particularly suitable for the narrow region, such as in the construction-dense urban area. The low-speed Maglev train has strong line adaptability.

3. Low construction and maintenance costs

The advantages of low-speed Maglev train, such as climbing ability, small turning radius, adaptability for complex terrain, and dense urban space, can reduce the demolition, which means reducing construction costs.

No contact friction of low-speed Maglev train with the track can reduce the maintenance costs of the vehicle and track.

Onboard computer system can realize the automatic vehicle monitoring and diagnostics, and the system adopts the modular maintenance, which can save maintenance workload and reduce maintenance costs.

4. Suspending and guiding function integration, less-board equipment, light weight body

Compared with the independent guidance system of TR Maglev train, HSST Maglev suspension and guidance function are integrated, which can reduce the number of onboard electromagnet and controllers.

5. Train can "hold track" to run, which means the good safety

The operation mode of train and track integration cannot result in the derailment and rollover accidents. Even if the power is cut, there is the onboard power to maintain the suspension until safely parking. The design of redundant components can make the low-speed Maglev train have a good safety performance. In addition, for the line design of low-speed Maglev train, low-speed Maglev line is suitable for overhead and has relatively higher security than the underground line.

## 1.8 Conclusion

For low-speed Maglev train, the Japanese HSST system has achieved commercial operation. South Korea Incheon International Airport is also currently building a 6.1-km commercial model line, and China is also actively discussing the issue of the construction of low-speed Maglev business lines. These phenomena show that the low-speed Maglev transportation as a new urban transport has a good prospect. Therefore, future research aiming at low-speed Maglev train will focus on the aspects of engineering applications. For the suspension control system, to improve the reliability and fault tolerance is the goal of Maglev suspension control system. In order to reduce the project cost, avoiding the problems of static floating vehicle–track coupling self-excited vibration of low-speed Maglev system is the future research requirements.

High-speed Maglev train technology has tremendous advantages. Although high-speed Maglev train has successfully conducted the commercial trial, it is very necessary through the growing up trunk of operational test. If the basic suspension control issues have been resolved, the high-speed Maglev research could proceed from the overall. Through the analysis of the vertical and lateral vehicle dynamics modeling, parameter identification problems, and considering various boundary conditions, the design of the suspension and guidance system can be optimized. The coupled vibration problems of vehicle–track in the suspension and guide system of high-speed Maglev train should also be investigated, which can solve the vehicle–track coupling vibration phenomena in the large-span beam orbit. In addition, some research work should be carried out, such as traction control, operation control, and high-precision positioning means of synchronous traction. Meanwhile, as an innovation in the field of high-speed Maglev train, though the electromagnetic hybrid permanent magnet technology of high-speed Maglev train has been verified, much research work should be carried out, such as safety and reliability of the suspension system and other engineering application problems.

Currently, Germany's TR series high-speed Maglev train and Japan's HSST series low-speed Maglev train have been put into commercial operation and shown great advantages of Maglev transportation. But there are also some problems.

LIM is adopted to realize the traction of electromagnetic low-speed Maglev train, which has low traction efficiency and low energy efficiency. When the maximum design speed reaches 110 km/h, the traction efficiency is only 70 %, and the efficiency will decrease with decreasing speed. The long stator synchronous traction is adopted for TR08 Maglev train, and the traction efficiency can reach 90 %. Hence, the traction efficiency of HSST is low and energy efficiency is low. In addition, considering that the traction motor primary coil of HSST is the short stator coil, the end effects will occur, which shows the train running resistance. With the increase of train speed, the train acceleration capability will drastically decrease. The maximum train speed of the hill line is only 100 km/h.

The redundancy of high-speed electromagnetic Maglev train is very strong, but its structure is complex and difficult to control the vehicle. For example, the independent guidance system is adopted and the guidance system of single vehicle is about 6 t. In addition, the turning radius of high-speed Maglev train is big, and the line choice is difficult, which is not suitable to the city traffic and the application scope is limited.

In order to solve these problems, some innovative research work and new Maglev transportation system structures about Maglev technologies need to be carried out in future.

## References


1. Yan LG. Progress of high-speed Maglev in China [J]. IEEE Trans Appl Supercond. 2002;12:944–7.
2. Zhang RH, Yan LG, Xu SG, et al. Comparison of several projects of high speed Maglev [J]. Adv Technol Electr Eng Energy. 2004;23(2):46–50.
3. Wu Xiangming. Maglev Train [M]. Shanghai: Shanghai Science and Technology Press; 2003.
4. Scott D, Free J. Maglev: how they're getting trains off the ground [R]. Popular Science, 1973. p. 135.
5. He JL, Rote DM, Coffey HT. Study of Japanese electrodynamic-suspension Maglev systems [R]. Technical report. 1994-04-01.
6. Zhang Q, Lin F, You XJ, et al. A novel stator section crossing method of long stator linear synchronous motor for Maglev vehicles [C]. IPEMC 2006, Shanghai, China.
7. Chang Wensen. Maglev technology development and automatic control [C]. Yichang: 22th Chinese control conference. 2003;8:27–30.
8. Sinha PK. Electromagnetically suspension dynamics control [M]. London: Peter Peregrinus Ltd; 1987.
9. Najafi F, Nassar F. Comparison of high-speed rail and Maglev systems [J]. J Transp Eng. 1996;122(4):276–81.
10. Yasuda Y, Fujino M, Tanaka M, et al. The first HSST maglev commercial train in Japan [C]. In: Proceedings of the 18th international conference on magnetically levitated systems and linear drives (MAGLEV 2004). 2004. p. 76–85.
11. Schweitzer G, Maslen EH. Magnetic bearings theory, design, and application to rotating machinery [M]. Berlin/Heidelberg: Springer; 2009.
12. Akira Chiba, Tadashi Fukao, Osamu Ichikawa, Masahide Oshima, Masatsugu Takemoto, Dorrell DG. Magnetic bearings, theory and applications [M]. Oxford: Elsevier Press; 2005.
13. Wei Q. C., Kong Y. J., Shi Jin, System and Technology for Maglev Transit [M]. China Science and Technology Press; 2010.
14. Liu Huaqing, Li Zhiye. German maglev trains [M]. Chengdu: University of Electronic Science & Technology Press; 1995. p. 17–30.
15. Wei Jichao. Maglev railway system and technology [M]. Beijing: China Science & Technology Press; 2003.
16. Yoshihide Yasuda, Masaaki Fujino, Masao Tanaka, et al. The first HSST maglev commercial train in Japan [C]. In: Maglev'2004 proceedings. 2004;1:76–85.
17. Ren Zhongyou, Wang Suyu, Wang Jiasu. Levitation force of multi-block YBaCuO high temperature superconductors over a permanent magnetic railway [J]. Cryogen Supercond. 2000;28(2):17–21.
18. Noriyuki Shirakuni, Motoaki Terai, Katsutoshi Watanabe. The status of development and running tests of Superconducting Maglev [C]. In: Maglev 2006 proceedings. 2006. p. 4–9.
19. Sam Gurol, Bob Baldi. General atomics urban Maglev program status [C]. In: Maglev' 2006 proceedings. 2006. p. 44–47.
20. Richard Thornton, Tracy Clark, Brian Perreault, Jim Wieler, Steve Levine. An M3 Maglev system for old dominion university [C]. In: Maglev 2008 proceedings. 2008;1:29–41.
21. Zhang Kunlun. Maglev system and its control [D]. Chengdu: Southwest Jiaotong University; 1990.

22. Shun Zhang. The low medium speed train pilot system past pilot accreditation [J]. Res Urban Rail Transp. 2001;41(4):69–73.
23. Liu Zhiming. Low medium speed maglev train technology and domestic engineering development [J]. Rail Transp. 2006;46(4):46–50.
24. Shi Xiaohong, She Longhua, Chang Wensen. The bifurcation analysis of the EMS maglev vehicle-coupled-guideway system [J]. Chin J Theor Appl Mech. 2004;36(5):634–40.
25. Wang Hongpo. Vehicle-guideway dynamic interaction of the EMS low speed maglev vehicle [D]. Changsha: University of Defense Technology; 2007.
26. Zhao Chunfa. Maglev vehicle system dynamics [D]. PhD thesis. Chengdu: Southwest Jiaotong University; 2004.
27. Zhai Wanming, Zhao Chunfa, Cai Chengbiao. On the comparison of dynamic effects on bridges of maglev trains with high-speed wheel/rail trains [J]. J Traffic Transp Eng. 2001;1(1):7–12.
28. Zhou Youhe, Wu Jianjun, Zheng Xiaojing. Analysis of dynamic stability for magnetic levitation vehicles by lyapunov characteristic number [J]. Acta Mech SINICA. 2000;32(1):42–51.
29. Li Yungang, Chang Wensen. Cascade control of an EMS maglev vehicle's levitation control system [J]. Acta Autom SINICA. 1999;25(2):247–51.
30. Cui Peng, Li Jie, Zhang Kun. Design of the suspension controller based on compensated feedback linearization [J]. J China Rail Soc. 2010;32(2):37–40.
31. Long Zhiqiang, Chen Huixing, Chang Wensen. Fault tolerant control on single suspension module of maglev train with electromagnet failure [J]. Control Theor Appl. 2007;24(6):1033–7.
32. Long Zhiqiang, Cai Ying, Xu Xin. Comprehensive fault evaluation on maglev train based on estimation of distribution algorithms [J]. Control Decis. 2009;24(4):551–6.
33. Li Yun, Li Jie, Zhang Geng, Tian Wen-jing. Disturbance decoupled fault diagnosis for sensor fault of maglev suspension system [J]. J Cent S Univ Technol. 2013;20(6):1545–51.
34. Li Yun, He Guang, Li Jie. Nonlinear Robust observer-based fault detection for networked suspension control system of Maglev train [J]. Math Probl Eng 2013;(2013), Article ID 713560:1-7.
35. Long Zhiqiang, Xue Song, He Guang, Xie Yunde. Fault-diagnosis for the accelerometer of maglev suspension system based on signal comparison [J]. Chin J Sci Instrum. 2011;32(12):2642–7.
36. Long Zhiqiang, Zhou Xiaobin, Yang Quanlin, Yin Liming. System study of permanent maglev train [J]. Electr Drive Locomot. 1996;3:8–11.
37. Li Hong, Zuo Peng, Liu Weizhi, Yuan Weici. Study on the 6t single bogie maglev test car [J]. J China Rail Soc. 1999;21(2):26–9.
38. Hong Huajie. Study on train-rail coupling vibration of EMS low-speed maglev vehicle [D]. PhD thesis. Changsha: University of Defense Technology; 2005.
39. Zhou DF, Li J, Hansen CH. Suppression of the stationary Maglev vehicle-bridge coupled resonance using a tuned mass damper [J]. J Vib Control. 2013;19(2):191–203.
40. Zhou DF, Hansen CH, Li J. Suppression of Maglev vehicle–girder self–excited vibration using a virtual tuned mass damper [J]. J Sound Vib. 2011;330:883–901.
41. Lee N, Han H, Lee J, et al. Maglev vehicle/guideway dynamic interaction based on vibrational experiment [C]. In: Proceedings of international conference on electrical machines and systems, 2007. p. 1999–2003.
42. Teng YF, Teng NG, Kou XJ. Vibration analysis of Maglev three-span continuous guideway considering control system [J]. J Zhejiang Univ Sci A. 2008;9(1):8–14.
43. Long Zhiqiang, Gai Ying, Xu Xin. Comprehensive fault evaluation on maglev train based on estimation of distribution algorithms [J]. Control Decis. 2009;24(4):551–556.
44. Li Yun, He Guang, Li. Jie. Nonlinear robust observer-based fault detection for networked suspension control system of Maglev train. Math Probl Eng 2013;(2013), Article ID 713560.
45. Xue Song, Long Zhiqiang, He Ning, Chang Wensen. A high precision position sensor design and its signal processing algorithm for maglev train [J]. Sensors. 2012;12:5225–45.

46. Zhang Zhizhou. Magnet design and signal processing technology study of permanent magnet electromagnetic low-speed maglev train's suspension system [D]. PhD thesis. Changsha: University of Defense Technology; 2011.
47. Li Yungang, Cheng Hu, Long Zhiqiang. Stability analysis and controller design of hybrid EMS Maglev system [C]. In: The 8th international symposium on magnetic suspension technology. 2005. p. 74–78.
48. Chen Huixing. Design and control of magnets in permanent EMS maglev trains [D]. PhD thesis. Changsha: University of Defense Technology; 2009.
49. Li Lu, Zhou Wenwu, Wu Jun. Temperature drift compensation for gap sensor of high speed maglev train [J]. Chin J Sensors Actuators. 2008;21(1):70–3.
50. Hao Aming. Study on key technology of ordinary high speed maglev train guidance system [D]. Changsha: University of Defense Technology; 2008.
51. Li Lu, Zhou Wenwu, Wu Jun. Research on HSMT levitation gap signal processing [J]. Chin J Sci Instrum. 2008;29(9):2001–4.
52. Guo Xiaozhou, Wang Ying, Wang Shixiong. Location and speed detection system for high-speed maglev vehicle [J]. J Southwest Jiaotong Univ. 2004;39(4):455–9.
53. Qian Cunyuan, Han Zhengzhi, Shao Derong, Xie Weida. Study of rotor position detection method for linear synchronous motor [J]. Proc CSEE. 2006;26(15):129–33.
54. Li Lu, Wu Jun, Luo Honghao. Research of joint-passing of speed and position detection system of maglev train [J]. J China Rail Soc. 2009;31(2):69–72.
55. Dai Chunhui, Long Zhiqiang, Xie Yunde, Xue Song. Research on the filtering algorithm in speed and position detection of maglev trains [J]. Sensors. 2011;11:7204–18.

# Chapter 2
# Technology Development and Application Research of Maglev Control

Suspension control, traction control, and operation control technologies are three key technologies of the electromagnetic suspension Maglev train [1]. Traction control and operational control in the wheel-rail transportation system have been fully studied and applied. Compared with the suspension control technology, the traction control and operation control technologies have relatively matured [2–4]. Since the Maglev suspension control technology is unique, it is still very necessary to study it in engineering applications [5, 6].

## 2.1 Introduction

In essence, the electromagnetic suspension system is unstable. Hence, the active suspension control must be applied to ensure the stability of trains and realize the train operation without mechanical contact. In addition, the suspension system performance directly influences the safety and comfort of Maglev train. Meanwhile, due to the complexity of the controlled object and the uncertainty of operation environment, the suspension control technology is always the difficulty and the core issue of the EMS Maglev train [7–9].

The suspension control system mainly comprises suspension sensors, suspension controllers, suspension choppers, supply power, and other auxiliary equipment. The controllers are the core of the suspension control system [10–12]. The suspension control principle is shown in Fig. 2.1.

The suspension sensors can measure the gap between the electromagnet and the track, and the current and motion state of electromagnets, which are fed back to the suspension controller. The suspension controllers can compare the measured value with the set value of levitation gap. If the measured value is greater than the set value, the control quantities are calculated and output based on the control law. The control voltage across the electromagnet is increased through the suspension



Z. Liu et al., *Maglev Trains*, Springer Tracts in Mechanical Engineering,
DOI 10.1007/978-3-662-45673-6_2

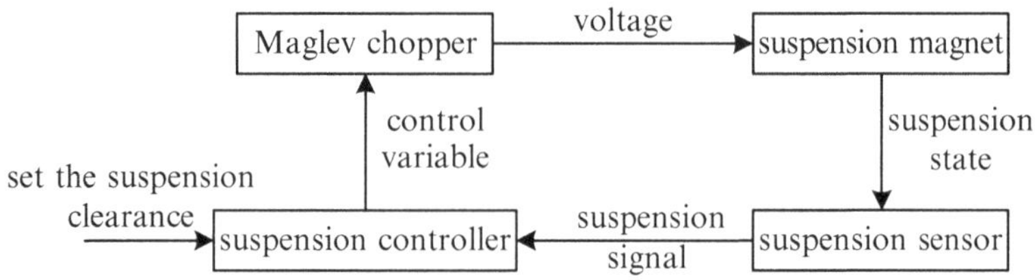


**Fig. 2.1** Control structure of suspension system

chopper, and the suspension current is added to increase the electromagnetic force, which can make the electromagnet upward move and reduce the suspension gap. Otherwise the control voltage is decreased to reduce the electromagnetic force, so that the downward movement of the electromagnet is carried out and the suspension gap is increased. In addition, in order to improve the performance of suspension system, the suspension controller can provide the damping for the suspension system by the speed feedback. The current response speed of the electromagnet can be improved through the current response [13–15].

The suspension system structure of Maglev train is complex and the model is strongly nonlinear. In addition, due to the external disturbances and parameter perturbation, the system model has greater uncertainty. The suspension control system must ensure the stability of the Maglev train under various conditions, which is a basic requirement of Maglev technologies.

## 2.2 Suspension Control Plan of Maglev Train

The low-speed EMS Maglev train uses the hierarchical structure based on the "module" concept, and the module is an independent unit that integrates the suspension, guidance, and traction functions [16–18]. The unit interior contains four suspension electromagnets, which are divided into two groups and formed two suspension endpoints through the interaction with the track. Two modules form a bogie by the connection of anti-roll beams. Several bogies form the suspension system of Maglev train through the interaction between electromagnetic force and track. For the suspension control technology research, we can choose a suspended endpoint, a module, or Maglev train as the controlled object. The following several control schemes can be presented, namely, single electromagnet suspension control scheme, suspension control module scheme, centralized control scheme, and overlapping suspension control scheme [19–21].

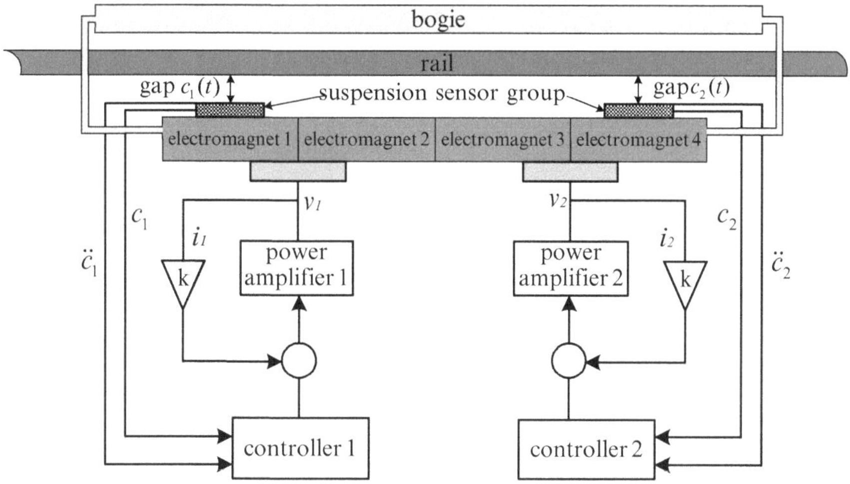


**Fig. 2.2** The structure of single point suspension control system

### 2.2.1 Single Electromagnet Suspension Control Scheme

Two group electromagnets inside the module are equivalent to two completely independent single electromagnets, so the module can be simply divided into two separately controlled objects as shown in Fig. 2.2. Since the module is a rigid body, the states of two suspension end point will affect each other, particularly when the state of a suspension endpoint is adjusted due to over rail joints or disturbed.

For the suspension control scheme of single electromagnet, this kind of interference is considered as the outside interference of system and can be passively suppressed by increasing the robustness of the control algorithms. At present, most of the suspension control algorithms are based on the control scheme design and analysis of single electromagnet suspension.

Using two power amplifiers, two controllers can control four electromagnets, and each controller is corresponding to two series connection electromagnet. Controller 1 only receives the signals from the left sensor groups, and controller 2 only receives the signals from the right sensor groups, namely, the system is divided into two completely separate controlled subsystems. The method only requires fewer components and lower cost, but it is affected by the coupled [22–25].

### 2.2.2 Module Suspension Control Scheme

For the state of the module suspension system coupled problem, through the decoupled control algorithm design, the module is considered as the controlled

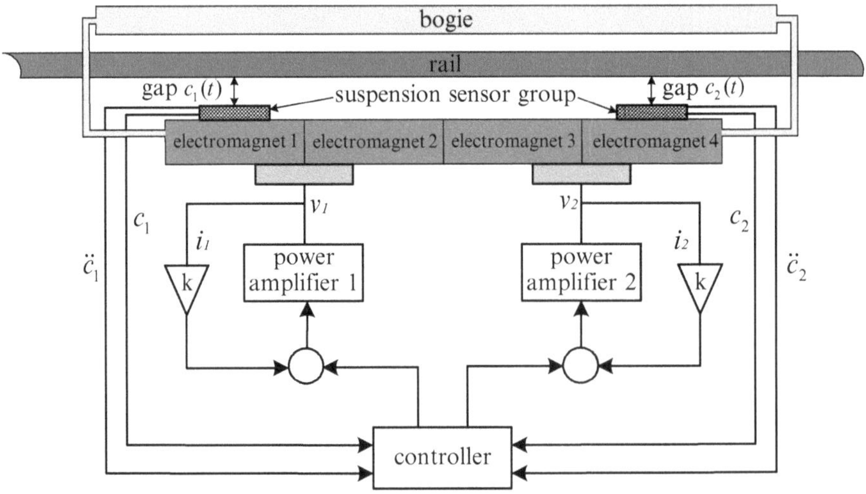


**Fig. 2.3** The structure of module suspension control system

object, which can actively suppress the mutual coupling between the internal modules of two suspended endpoints [26–28]. The way can improve the suspension system performance without increasing system cost, which is shown in Fig. 2.3.

Only one controller in the entire module is used to control two pairs of electromagnets. The corresponding electromagnets for each pair electromagnets are connected in series. The plan requires a minimum of component parts and is the most economical plan. The controller can receive the signals from the sensors of two end points in the module. It is unnecessary to divide the system in the plan, but the whole module as a complete system is decoupled. The plan that combines the information at both ends of the sensor can effectively eliminate the influence of mechanical coupling ends of the module. The controller uses a redundant structure to improve system reliability. However, any electromagnet or amplifier controller failure will lead to the suspension failure.

### 2.2.3 *Concentrated Suspension Control Scheme*

A bogie or Maglev train is considered as a controlled object; a controller is adopted to complete its suspension control. The centralized control structure requires a mathematical model of a bogie or Maglev train system, and the comprehensive multiple suspension points to obtain the status information of controlled objects with the sensor signals, which results in the complex structure of suspension control system and the difficult control algorithm design.

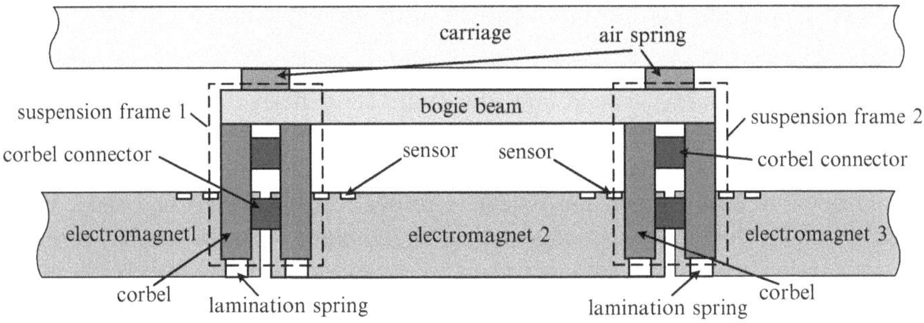


**Fig. 2.4** Overlapping suspension control scheme

### *2.2.4 Overlapping Suspension Control Scheme*

In order to ensure the redundancy of suspension system, the overlapping suspension control scheme can be adopted for high-speed Maglev train. Each suspension frame has two suspension arms, and the two suspension arms via an upper connection component and a lower connection component are connected by a bolt, which can form the overall stiffness and elasticity. The suspension electromagnets are mounted on the frame between two adjacent suspension frames. So, two suspension arms of a suspension frame are corresponding to different suspension electromagnets, and they can form the overlapping suspension structure. In two suspension units of the overlapping structure, two sensors are respectively equipped with, which can make the suspension system have redundancy capabilities when pass the rail joints as shown in Fig. 2.4. If one side suspension electromagnets of the overlapping structure cannot work due to the fault, the other one is still able to take the suspension function of suspension frame by the overlapping structure to avoid the falling accident.

Considering that the overlapping structure makes high-speed Maglev train have the huge lower substructure, the system's weight and complexity and the difficulty of control will increase.

## 2.3 Suspension Control Algorithms of Maglev Train

### *2.3.1 Basic Requirements of Suspension Control Algorithms*

The study on suspended control algorithms of Maglev train is one of the most active topics in the Maglev technologies. Based on the structural characteristics and engineering requirements of suspension system in EMS Maglev train, some basic requirements on suspension control algorithms are listed as follows [29, 30]:

1. Good adaptability for the uncertainty of suspension system models

Whether single electromagnet or module suspension control schemes are adopted, each Maglev vehicle contains multiple suspension objects. There are large or small differences between the parameters of these objects each other, and there are also differences between the train and the train's suspension system parameters. If each suspension object is respectively designed and the control parameters are debugged, the industrialization requirements are obviously meet for Maglev train. In addition, the changes of work environment can also result in the suspension system parameters perturbation. Therefore, the suspension controller should have good adaptability for the module differences and system parameter perturbation caused by the environmental change.

2. Good robustness for the external interference of suspension system

In the operation process of suspension system, the suspension magnetic field, electromagnetic radiation of chopper, track irregularity, and other factors can cause the state fluctuation of the sensor measured signals. The change of the passenger number can also result in the load change of suspension system. Therefore, the suspension controllers should have good robustness for these two external disturbances.

3. Good suppression ability for the track coupled vibration

The suspension system consists of suspension electromagnet and track suspension. The suspension is the result of their interaction. In the laboratory environment, if the track stiffness coefficient is large enough, the track can be usually considered as a rigid body, and the elastic vibration and geometric distortion of tracks are ignored. In the actual line, the track stiffness cannot be very large. In the operation process of suspension system, the resonance between the vehicle and the track easily occurs and even causes the system to diverge. All of Germany's TR-04 system, HSST-04 system in Japan, and the US AMT systems have the suspension system failure due to vehicle vibrations of the vehicle–track coupling. Therefore, whether considering the operating safety or comfort of Maglev train, the suspension controller should have good vibration suppression for the vehicle–track coupling vibration under different conditions.

### 2.3.2 *Suspension Control Methods of Maglev Train*

In the early development of Maglev train, due to the constraints of control theory and hardware conditions, the nonlinear model of suspension system is usually expanded through Taylor series near the small neighborhood zone of work point, and the higher-order infinitesimal terms are ignored to obtain the system linear approximation model. Then the classical control theory algorithm is adopted to design the suspension system of Maglev train. The following are two disadvantages for the method:

1. The controller based on the linear approximation model has good control performance around the set work point. However, when the system is set away from the work point, the system performance will deteriorate sharply, and even the instability will occur.
2. Due to the suspension system model uncertainty, the performance of control algorithms based on the classical control theory is unsatisfactory to suppress the parameter perturbation, external interference, and vehicle–track coupling vibrations of the system.

In order to solve the first problem, the gain adjustment control method is proposed. In the operation range of suspension system, more than one work points are selected, and the multiple linear models can be obtained through the system model linearization in the neighborhood zone of each work point with Taylor series expansion method. In addition, the reasonable gain control is respectively designed for these linear models. The suspension control system can automatically select the appropriate control gain based on the current suspension state.

With the development of nonlinear science, the feedback linearization of nonlinear systems has become more mature linearization method. Utilizing the principle of differential geometry, through rigorous mathematical derivation, the linear model of nonlinear system can be obtained by using nonlinear term exact offset method. The linear model obtained in this way is not only accurate, but also global, namely, does not depend on the set work point [31–33]. Considering that the feedback linearization procedure contains multiplication and square root operations, the hardware system requires high performance demand. With the development and popularization of digital controllers, the feedback linearization algorithm, which can achieve more convenient and accurate performance, has been widely used gradually in the design of the suspension control system [34–36].

With the rapid development of nonlinear system theory, modern control theory, and microelectronics technology, in order to solve the second disadvantage, more and more modern control methods and nonlinear control methods are applied to the design of the suspension control algorithm, such as optimal control, fuzzy control, neural network control, adaptive control, robust control, sliding mode control, and so on [37, 38].

These control methods are presented as follows [39–43]:

1. Optimal control method

With DSP controller, an adaptive suspension controller optimization method based on model reference is realized, which can have good inhibitory for the external force interfere and the suspended mass change in ensuring the premise tracking performance.

2. $\mu$ analysis/synthesis method

$\mu$ analysis/synthesis method can be adopted in the suspension control. Based on the different idealized assumptions for actual system, four structure models are derived. For different models, the maximum, minimum, and nominal values of the parameters are determined by means of experiments and measurements. The model with the simplest structure and the nominal value parameter in the four model structures is selected as the nominal model. The additive uncertainty is calculated, and the robust performance indicator is established to determine the frequency weight function uncertainty. The algorithm can efficiently suppress the model parameter perturbation, but the choice of the nominal model is more difficult, and the realization of control algorithm is more complex.

3. Robust control method

For the cascade control method of suspension system, the suspension system is decomposed into the two subsystems with current loop and position loop [29]. The robust theory is adopted to design current loop, which can well compensate the great delay caused by electromagnetic inductance. In addition, classical control theory is adopted to design the suspension controller to achieve a full-size bogie suspension.

Since the position loop is designed with classical control theory, the suppression performance for the model perturbation, external interference, and vehicle–track coupling vibration is not obvious. But the cascade control idea has generated a great deal of influence for the suspension control algorithms design.

4. Sliding mode control method

The dynamic sliding mode suspension controller can efficiently suppress the system parameter perturbation and external interference and to some extent can weaken the chattering in the static variable structure controller [44–46].

In addition, based on suspension clearance and electromagnet acceleration signals, through the obtained damping signals from robust state observer, the stable single bogie suspension of Maglev train can be realized with the design suspension controller, which can solve the vehicle–track vibration problem of Maglev train to some extent [47–49].

At present, there are many new suspension control algorithms. Some algorithms have been applied in the laboratory experiment verification, and more still remains in the simulation stage [50–52]. The algorithm effectiveness still needs to be verified in practical engineering applications [53–59].

## 2.4 Conclusion

Maglev control technology has been widely used and rapidly developed. In this chapter, from the applications of low-speed Maglev train and high-speed Maglev train, the applications for Maglev control technical analysis are introduced [60–63]. In addition, Maglev control technology involves a number of research areas. The summary is not comprehensive and complete. It is hoped to provide some help for Maglev control technology research.

## References


1. Chang Wensen. Maglev technology development and automatic control [C]. Yichang: 22th Chinese control conference. 2003;8:27–30.
2. Sinha PK. Electronmagnetically suspension dynamics control [M]. London: Peter Peregrinus Ltd; 1987.
3. Najafi F, Nassar F. Comparison of high-speed rail and Maglev systems [J]. J Transp Eng. 1996;122(4):276–81.
4. Yasuda Y, Fujino M, Tanaka M, et al. The first HSST maglev commercial train in Japan [C]. In: Proceedings of the 18th international conference on magnetically levitated systems and linear drives (MAGLEV 2004). 2004. p. 76–85.
5. Gerhard Schweitzer, Maslen EH. Magnetic bearings theory, design, and application to rotating machinery [M]. Berlin/Heidelberg: Springer; 2009.
6. Akira Chiba, Tadashi Fukao, Osamu Ichikawa, Masahide Oshima, Masatsugu Takemoto, Dorrell DG. Magnetic bearings, theory and applications [M]. Oxford: Elsevier Press; 2005.
7. Chang Wensen. The latest research trends of maglev trains [J]. Chin J Sci. 1993;33(6):34–6.
8. Wu Xiangming. Maglev trains [M]. Shanghai: Shanghai Science & Technology Press; 2003.
9. Liu Huaqing, Li Zhiye. German maglev trains [M]. Chengdu: University of Electronic Science & Technology Press; 1995. p. 17–30.
10. Wei Jichao. Maglev railway system and technology [M]. Beijing: China Science & Technology Press; 2003.
11. Yoshihide Yasuda, Masaaki Fujino, Masao Tanaka, et al. The first HSST maglev commercial train in Japan [C]. In: Maglev 2004 proceedings. 2004;1:76–85.
12. Ren Zhongyou, Wang Suyu, Wang Jiasu. Levitation force of multi-block YBaCuO high temperature superconductors over a permanent magnetic railway [J]. Cryog Supercond. 2000;28(2):17–21.
13. Noriyuki Shirakuni, Motoaki Terai, Katsutoshi Watanabe. The status of development and running tests of Superconducting Maglev [C]. In: Maglev 2006 proceedings. 2006. p. 4–9.
14. Sam Gurol, Bob Baldi. General atomics urban Maglev program status [C]. In: Maglev 2006 proceedings. 2006. p. 44–7.
15. Richard Thornton, Tracy Clark, Brian Perreault, Jim Wieler, Steve Levine. An M3 Maglev system for old dominion university [C]. In: Maglev 2008 proceedings. 2008;1:29–41.
16. Zhang Kunlun. Maglev system and its control [D]. PhD thesis. Chengdu: Southwest Jiaotong University; 1990.
17. Shun Zhang. The low medium speed train pilot system past pilot accreditation [J]. Res Urban Rail Transp. 2001;41(4):69–73.
18. Liu Zhiming. Low medium speed maglev train technology and domestic engineering development [J]. Rail Transp. 2006;46(4):46–50.
19. Hong Huajie. Vehicle-guideway coupled vibration of EMS type low speed maglev trains [D]. PhD thesis. Changsha: University of Defense Technology; 2005.
20. Shi Xiaohong, She Longhua, Chang Wensen. The bifurcation analysis of the EMS maglev vehicle-coupled-guideway system [J]. Chin J Theor Appl Mech. 2004;36(5):634–40.
21. Wang Hongpo. Vehicle-guideway dynamic interaction of the EMS low speed maglev vehicle [D]. Changsha: University of Defense Technology; 2007.
22. Zhao Chunfa. Maglev vehicle system dynamics [D]. PhD thesis. Chengdu: Southwest Jiaotong University; 2004.
23. Wanming Zhai, Zhao Chunfa, Cai Chengbiao. On the comparison of dynamic effects on bridges of maglev trains with high-speed wheel/rail trains [J]. J Traffic Transp Eng. 2001;1(1):7–12.
24. Zhou Youhe, Wu Jianjun, Zheng Xiaojing. Analysis of dynamic stability for magnetic levitation vehicles by Lyapunov characteristic number [J]. Acta Mech Sinica. 2000;32(1):42–51.
25. Teng YF, Teng NG, Kou XJ. Vibration analysis of Maglev three-span continuous guideway considering control system [J]. J Zhejiang Univ Sci A. 2008;9(1):8–14.

26. Zhou DF, Li J, Hansen CH. Suppression of the stationary Maglev vehicle-bridge coupled resonance using a tuned mass damper [J]. J Vib Control. 2013;19(2):191–203.
27. Zhou DF, Hansen CH, Li J. Suppression of Maglev vehicle-girder self-excited vibration using a virtual tuned mass damper [J]. J Sound Vib. 2011;330:883–901.
28. Lee N, Han H, Lee J, et al. Maglev vehicle/guideway dynamic interaction based on vibrational experiment [C]. In: Proceedings of international conference on electrical machines and systems. 2007. p. 1999–2003.
29. Li Yungang, Chang Wensen. Cascade control of an EMS maglev vehicle's levitation control system [J]. Acta Autom Sinica. 1999;25(2):247–51.
30. Cui Peng, Li Jie, Zhang Kun. Design of the suspension controller based on compensated feedback linearization [J]. J China Railw Soc. 2010;32(2):37–40.
31. Long Zhiqiang, Xue Song, He Guang, Xie Yunde. Fault-diagnosis for the accelerometer of maglev suspension system based on signal comparison [J]. Chin J Sci Instrum. 2011;32(12):2642–7.
32. Long Zhiqiang, Zhou Xiaobing, Yang Quanlin, Yin Liming. System study of permanent maglev train [J]. Electr Drive Locomot. 1996;3:8–11.
33. Li Hong, Zuo Peng, Liu Weizhi, Yuan Weici. Study on the 6 t single bogie maglev test car [J]. J China Railw Soc. 1999;21(2):26–9.
34. Shi Xiaohong, She Longhua, Chang Wensen. The bifurcation analysis of the EMS maglev vehicle-coupled-guideway system [J]. Acta Mech Sinica. 2004;36(5):634–40.
35. Long Zhiqiang, Chen Huixing, Chang Wensen. Fault tolerant control on single suspension module of maglev train with electromagnet failure [J]. Control Theory Appl. 2007;24(6):1033–7.
36. Long Zhiqiang, Gai Ying, Xu Xin. Comprehensive fault evaluation on maglev train based on estimation of distribution algorithms [J]. Control Decis. 2009;24(4):551–6.
37. Li Yun, Li Jie, Zhang Geng, Tian Wenjing. Disturbance decoupled fault diagnosis for sensor fault of maglev suspension system [J]. J Cent S Univ Technol. 2013;20:1545–51.
38. Li Yun, He Guang, Li Jie. Nonlinear robust observer-based fault detection for networked suspension control system of Maglev train [J]. Math Probl Eng. 2013; Article ID 713560:1–7.
39. Xue Song, Long Zhiqiang, He Ning, Chang Wensen. A high precision position sensor design and its signal processing algorithm for Maglev train [J]. Sensors. 2012;12:5225–45.
40. Zhang Zhizhou. Magnet design and signal processing technology study of permanent magnet electromagnetic low-speed maglev train's suspension system [D]. PhD thesis. Changsha: University of Defense Technology; 2011.
41. Li Yungang, Cheng Hu, Long Zhiqiang. Stability analysis and controller design of hybrid EMS Maglev system [C]. In: The 8th international symposium on magnetic suspension technology. 2005. p. 74–8.
42. Chen Huixing. Design and control of magnets in permanent EMS maglev trains [D]. PhD thesis. Changsha: University of Defense Technology; 2009.
43. Li Lu, Wu Jun, Zhou Wenwu. Temperature drift compensation for gap sensor of high speed maglev train [J]. Chin J Sensors Actuators. 2008;21(1):70–3.
44. Hao Aming. Study on key technology of ordinary high speed maglev train guidance system [D]. PhD thesis. Changsha: University of Defense Technology; 2008.
45. Li Lu, Zhou Wenwu, Wu Jun. Research on HSMT levitation gap signal processing [J]. Chin J Sci Instrum. 2008;29(9):2001–4.
46. Guo Xiaozhou, Wang Ying, Wang Shixiong. Location and speed detection system for high-speed maglev vehicle [J]. J Southwest Jiaotong Univ. 2004;39(4):455–9.
47. Qian Cunyuan, Han Zhengzhi, Shao Derong, Xie Weida. Study of rotor position detection method for linear synchronous motor [J]. Proc CSEE. 2006;26(15):129–33.
48. Li Lu, Wu Jun, Luo Honghao. Research of joint-passing of speed and position detection system of maglev train [J]. J China Railw Soc. 2009;31(2):69–72.
49. Dai Chunhui, Long Zhiqiang, Xie Yunde, Xue Song. Research on the filtering algorithm in speed and position detection of maglev trains [J]. Sensors. 2011;11:7204–18.

50. Hu Yefa. Maglev support technology and development [C]. In: The 5th domestic conference on Maglev bearings. Changsha: China Maglev and Gas Levitation Technical Committee; 2013.
51. Su Wenjun, Sun Yanhua, Yu Lie. Rotor periodic vibration suppression for controlled magnetic levitation system [J]. J Xi'an Jiaotong Univ. 2010;44(7):55–8.
52. Tian Yongsheng, Sun Yanhua, Yu Lie. Dynamical and experimental researches of active magnetic bearing rotor systems for high-speed PM machines [J]. Proc CSEE. 2012; 32(9):116–23.
53. Fang Jiancheng. The research progress of high-precision and minimal vibrant magnetic inertial actuator [C]. In: The 5th Chinese symposium on magnetic bearing, Hunan, Changsha, 2013, August 26th-28th, Report of congress.
54. Wen Tong, Fang Jiancheng. A feedback linearization control for the nonlinear 5-DOF flywheel suspended by the permanent magnet biased hybrid magnetic bearings [J]. Acta Astronaut. 2012;79:131–9.
55. Wang Xi, Fang Jiancheng, Fan Yahong, Liu Hu, Wang Chune, Wen Tong, Sun Jinji. Thimble permanent-magnet-biased radial magnetic bearing for magnetically-suspended-flywheel [J]. J Mech Eng. 2011;47(14):171–83.
56. Wu Gang. Study on system design and control methods of hybrid magnetic bearing momentum flywheel [D]. PhD thesis. Changsha: University of Defense Technology; 2006.
57. Konstantinos Michail, Argyrios Zolotas, Roger Goodall, et al. Sensor optimization via $H\infty$ applied to a MAGLEV suspension system [C]. In: International conference on control, automation and systems, Prague, Czech Republic, 2008, July 25–27.
58. Sung HK, Cho HJ, Yoo MH, et al. Fault tolerant control of electromagnetic levitation system [C]. In: The 18th magnetically levitated system and linear drives conference. China; 2004.
59. Long Zhiqiang, Xue Song, Chen Huixing. Passive fault tolerant control for suspension system of Maglev train based on LMI [J]. Comput Simul. 2008;25(2):265–8.
60. Zeng Xueming. On electric control system for active magnetic bearing [D]. PhD thesis. Nanjing: Nanjing University of Aeronautics and Astronautics; 2002.
61. Xu Longxiang, Xiao Jili. Study on the high-speed and high-precision processing maglev spindle system: advanced manufacturing technology [M]. Beijing: China Science & Technology Press; 1997. p. 660–3.
62. Zhang Jiansheng. Study and applications on digital control technology and power amplifier in magnetic levitation support system [D]. PhD thesis. Shanghai: Shanghai University; 2006.
63. Zhang Gang, Sun Chang, Zhang Jian, Zhang Hailong, Jiang Dede. Design and experimental verification of axial permanent magnetic bearings [J]. Bearing. 2012;11:5–9.

# Chapter 3
# Modeling and Controller Design of Suspension System of Maglev Train

## 3.1 Introduction

The electromagnetic suspension-type (also known as attraction type) Maglev technology is one of the mature track transit technologies at present. By virtue of this technology, both Japan's low-speed Maglev Nagoya Linimo and China's high-speed Maglev Shanghai Pudong Airport Line succeed in realizing the commercial operation of Maglev traffic line. The electromagnet of electromagnetic suspension-type train is installed under the track. The suspension force is produced by mutual attraction of the electromagnet and the ferromagnetic component on the track. In order to keep stable suspension of the vehicle, the size of electromagnetic force is adjusted by controlling the field current of electromagnet. The electromagnetic force is equal to the train body's gravity under a certain suspension gap to keep stable suspension of the vehicle. When the electromagnetic suspension-type train cannot realize self-steadiness in the direction of suspension, a closed-loop automatic control system is required [1]. The suspension control system, whose performance can directly affect the stability and safety of train's running, is regarded as one of the emphases of study with regard to Maglev train. The researches in this chapter mainly include building the system model and designing the suspension controller algorithm and the filtering algorithm.

## 3.2 Building Single-Point Suspension Model

Both the electromagnetic suspension-type low-speed Maglev train and the high-speed Maglev train apply the way of multi-bogie mechanical decoupling. This structure can distribute the weight of whole train into many suspension control units, which plays very subtle influence on each other and has approximately independent control over each other, and realize the whole train suspension. Therefore, from the



Z. Liu et al., *Maglev Trains*, Springer Tracts in Mechanical Engineering,
DOI 10.1007/978-3-662-45673-6_3

perspective of controlling system design, it is allowed to build the "single-point" suspension model by neglecting the influence among the control units.

Maglev train, relying on electromagnetic force, attracts the train body to locate at a certain distance from the track and keep steady. However, the track is not ideal steel body because it may generate a certain deformation under the effect of electromagnetic force. Under some specific conditions, the electromagnetic force may produce continuous excitement on the track and finally result in the drastic periodical deformation of the track, namely, the track vibration. The vibration is the result of interaction between the vehicle and track. Therefore, it is called as the vehicle-coupled-track vibration [2]. In order to eliminate the phenomenon of vehicle-coupled-track vibration, it is required to consider the factor of track upon modeling to build the single-point suspension model including the track features.

A complete single-point suspension model includes electromagnet parts, connecting piece, air spring, carriage, and track. The track has a very complex structure and the electromagnet has many freedom degrees of motion including pitching, yawing, rolling, and heaving. Therefore, the whole system is a complex nonlinear coupling system. It is difficult to analyze the system and the further simplification is required to highlight the major problems before modeling. The simplification of model is mainly made based on the following several hypotheses:

1. Only the steel girder track is considered and simplified to be the supported beam.
2. The suspension control mainly focuses on the vertical motion of electromagnet.
   The vertical motion of electromagnet is considered. The main reason is only that the motion of electromagnet in other directions plays very subtle influence on the vertical motion that is bound by the structure.
3. The bogie and the carriage are connected with air spring whose stiffness is very small and has big elasticity. Therefore, how the carriage affects the dynamic performance of suspension electromagnet is ignored.

In this chapter, the elastic track model is firstly built, and then single-point suspension model is built below.

### 3.2.1 Elastic Track Model

Figure 3.1 is the schematic diagram of single-point suspension model,
where:

$x_0$ – end location of electromagnet
$z_m$ – vertical displacement of electromagnet
$z_G$ – vertical displacement of track
$m$ – mass of electromagnet
$F_e$ – electromagnetic force
$i$ – current in electromagnet coil
$u$ – control voltage at both ends of electromagnet
$R$ – equivalent resistance of electromagnet coil

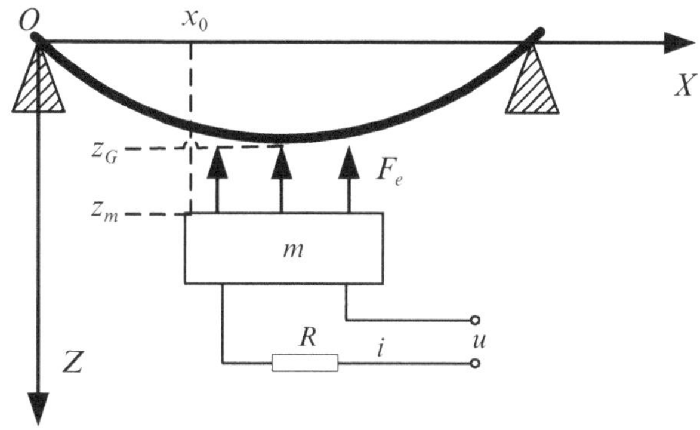


**Fig. 3.1** Single-point track coupling model

The track is simply supported beam and the coordinate $z_G\,(x, t)$ of any one point on the beam meets the Bernoulli–Euler dynamics equation:

$$E_G\, I_G \frac{\partial^4 z_G}{\partial x^4} - T_G \frac{\partial^2 z_G}{\partial x^2} + \rho_G \frac{\partial^2 z_G}{\partial t^2} + b_G \frac{\partial z_G}{\partial t} + k_G\, z_G = p\,(x,\ t) \tag{3.1}$$

In the formula:

$E_G$ – Young's elasticity modulus
$T_G$ – tension on the beam
$I_G$ – cross-sectional inertia of beam
$b_G$ – equivalent damping coefficient of beam
$\rho_G$ – mass linear density of beam
$k_G$ – equivalent elasticity coefficient of beam
$p(x, t)$ – distributed load density on the beam

Use the modal analysis method to order

$$Z_G\,(x,t) = \sum_n \phi_n(x) \cdot q_n(t) \quad n = 1, 2, \ldots, \infty \tag{3.2}$$

$\phi_n(x)$ and $q_n(t)$ in the formula, respectively, correspond to the $n$th-order modal and the generalized coordinates of beam. Set beam span of track, mass, as $l_G$, $m_G$, respectively.

$$\phi_n(x) = \sqrt{\frac{2}{m_G}} \sin\left(\frac{n\pi}{l_G}x\right) \tag{3.3}$$

$$\ddot{q}_n(t) + 2\eta_n\omega_n\dot{q}_n(t) + \omega_n^2 q_n(t) = Q_n \tag{3.4}$$

where $\eta_n$ represents the damping ratio of every order modal. This parameter should be decided by the factors such as materials, structure, and geometrical shape of the track. $\omega_n$ represents the inherent frequency of every order modal and meets the following conditions:

$$\omega_n = \left(\frac{n\pi}{l_G}\right)^2 \sqrt{\frac{E_G I_G}{\rho_G}} \tag{3.5}$$

$Q_n$ represents the generalized acting force of the $n$th order, and it is produced by the electromagnetic force which may be approximated at the distributed force as distributed uniformly on the upper surface of electromagnet. Set the effective length of single-point electromagnet as $l_e$. Calculate the generalized force as follows:

$$p(x,t) = \begin{cases} \frac{F_e}{l_e} & x_0 \le x \le x_0 + l_e \\ 0 & \text{the other section} \end{cases} \tag{3.6}$$

$$\begin{aligned} Q_n &= \frac{\int_0^{l_G} \phi_n(x) p(x,t)\,\mathrm{d}x}{\int_0^{l_G} (\phi_n(x))^2 \rho_G \mathrm{d}x} = \frac{\int_{x_0}^{x_0+l_e} \sqrt{\frac{2}{m_G}} \sin\left(\frac{n\pi}{l_G}x\right) \frac{F_e}{l_e}\mathrm{d}x}{\int_0^{l_G} \frac{2}{m_G}\sin^2\left(\frac{n\pi}{l_G}x\right)\rho_G \mathrm{d}x} \\ &= \frac{\sqrt{2m_G}\left(\cos\left(\frac{n\pi}{l_G}x_0\right) - \cos\left(\frac{n\pi}{l_G}(x_0 + l_e)\right)\right)}{n\rho_G \pi l_e} F_e \triangleq k_{qn} F_e \end{aligned} \tag{3.7}$$

For the actually available track, it is known from Formula (3.7) that with the increase of order, $k_{qn}$ will decrease rapidly. $Q_n$, the generalized acting force of high order, is very small in general. As the first order of generalized force plays the main role, all that requires considering is the first-order modal of track. Combining with the experiment phenomenon, the vibration frequency of the track for the vehicle-coupled-track vibration under the condition of equal track is mainly concentrated on some frequency band, which is generally regarded to be first order of vibration frequency of the track. It can be measured based on the experiments. The vibration frequency on the steel beam track is 50 Hz.

Suppose that the center of electromagnet be located in the middle of the track $x_0 = (l_G - l_e)/2$, with the known parameters as follows:

$$l_e = 1.5\ \text{m},\ l_G = 3\ \text{m},\ \rho_G = 1{,}000\ \text{kg/m},\ m_G = \rho_G \cdot l_G = 3{,}000\ \text{kg},$$
$$\eta_1 = 0.001$$

Calculate the first-order modal of track:

$$\begin{cases} \omega_1 = 2 \times 3.14 \times 50 = 314\ \text{rad/s} \\ Q_1 = \dfrac{\sqrt{2m_G}\left(\cos\left(\frac{\pi(l_G - l_e)}{2l_G}\right) - \cos\left(\frac{\pi(l_G + l_e)}{2l_G}\right)\right)}{\rho_G \pi l_e} F_e \\ \quad = 0.0232 \cdot F_e \triangleq k_{q1} \cdot F_e \\ \phi_1 = \sqrt{\dfrac{2}{m_G}} \sin\left(\dfrac{\pi(l_G - l_e)}{2l_G}\right) = 0.0182 \end{cases} \tag{3.8}$$

It can be obtained from Formula (3.2):

$$z_G = \phi_1 \cdot q_1 \quad \Rightarrow \quad q_1 = z_G/\phi_1 \tag{3.9}$$

Substitute Formulas (3.8) and (3.9) into Formula (3.4), and it can be obtained as follows:

$$\ddot{z}_G + 2\eta_1\omega_1\dot{z}_G + \omega_1^2 z_G = k_{q1}\phi_1 \cdot F_e \tag{3.10}$$

Formula (3.10) is the motion model of the track under the effect of electromagnetic force.

### 3.2.2 Single-Point Suspension Model

The single-point electromagnet model has been shown in Fig. 3.1, and the motion equation of electromagnet has been shown as follows:

$$m\ddot{z}_m = mg - F_e \tag{3.11}$$

Hence, the electromagnetic force $F_e$ is

$$F_e = \frac{\mu_0 N^2 A}{4} \cdot \frac{i^2}{z^2} \tag{3.12}$$

In the formula, $z$ represents the gap between the electromagnet and the track, $\mu_0$ represents the permeability of vacuum, $N$ represents the turns of electromagnet coil, and $A$ represents the effective polar area of electromagnetic coil.

Combining with $C = \frac{\mu_0 N^2 A}{4}$, the dynamics equation of electromagnet can be obtained from Formulas (3.11) and (3.12).

$$m\ddot{z}_m = mg - C \cdot \frac{i^2}{z^2} \tag{3.13}$$

The equation of electromagnet voltage and current is shown as follows:

$$u = R \cdot i + \frac{\mathrm{d}(L \cdot i)}{\mathrm{d}t} = R \cdot i + \frac{\mathrm{d}L}{\mathrm{d}t} \cdot i + \frac{\mathrm{d}i}{\mathrm{d}t} \cdot L \tag{3.14}$$

Therein $L$ represents the equivalent inductance of electromagnet coil and

$$L = \frac{2C}{z} \tag{3.15}$$

Substitute Formula (3.15) into (3.14); it can be obtained that

$$u = R \cdot i + \frac{2C}{z} \cdot \dot{i} - \frac{2C \cdot i}{z^2} \cdot \dot{z} \tag{3.16}$$

By combining Formulas (3.10), (3.13), and (3.16), the single-point track coupling system model can be obtained as follows:

$$\begin{cases} \ddot{z}_G + 2\eta_1\omega_1\dot{z}_G + \omega_1^2 z_G = k_{q1}\phi_1 \cdot F_e = k_{q1}\phi_1 C \cdot \frac{i^2}{z^2} \\ m\ddot{z}_m = mg - C \cdot \frac{i^2}{z^2} \\ z = z_m - z_G \\ u = R \cdot i + \frac{2C}{z} \cdot \dot{i} - \frac{2C \cdot i}{z^2} \cdot \dot{z} \end{cases} \tag{3.17}$$

The known system parameters are

$$C = 0.00545, \ m = 725 \text{ kg}, \ R = 4.4\Omega, \ g = 9.8 \text{ N/kg}$$

## 3.3 Design of Control System

The single-point suspension model shown in Formula (3.17) involves nonlinear items, very high-order and complicated direct control algorithm. And it cannot highlight the influence of parameters on the system performance. Therefore, the approximately separating design method is applied by combining the particularity of the system to design the controller in terms of electromagnet and track respectively [3, 4]. Finally, it can enable the whole performance to meet the requirements for the index. With respect to the index of controller design, it means that the system can be kept stable when the gap $z_e$ between electromagnet and track is 0.012 m. The electromagnet can keep fast trace of the track at the same time.

### *3.3.1 Analysis of Model Linearization and Open-Loop Stability*

Firstly, calculate the equilibrium point of the system as shown in Formula (3.18):

$$\begin{cases} z_e = 0.012 \text{ m} \\ i_e = \sqrt{\frac{mg}{C}} \cdot z_e \\ u_e = R \cdot i_e \end{cases} \tag{3.18}$$

Use Taylor expansion for linearization of Formula (3.18) at the equilibrium point to get the linearization model of the system:

$$\begin{cases} \ddot{z}_G = \left(2k_{q1}\phi_1 C\frac{i_e^2}{z_e^3} - \omega_1^2\right)\cdot z_G - 2\eta_1\omega_1\cdot\dot{z}_G - 2k_{q1}\phi_1 C\frac{i_e^2}{z_e^3}\cdot z_m + 2k_{q1}\phi_1 C\frac{i_e}{z_e^2}\cdot i \\ \ddot{z}_m = \frac{2Ci_e^2}{mz_e^3}\cdot z_m - \frac{2Ci_e^2}{mz_e^3}\cdot z_G - \frac{2Ci_e}{mz_e^2}\cdot i \\ \dot{i} = \frac{i_e}{z_e}\cdot\dot{z}_m - \frac{i_e}{z_e}\cdot\dot{z}_G - \frac{Rz_e}{2C}\cdot i + \frac{z_e}{2C}\cdot u \end{cases} \tag{3.19}$$

Select the state variable

$$X_S = \begin{bmatrix} z_m & \dot{z}_m & z_G & \dot{z}_G & i \end{bmatrix}^{\mathrm{T}} \tag{3.20}$$

Get the state space expression of single-point suspension model:

$$\begin{aligned} \dot{X}_S &= A_S\cdot X_S + B_S\cdot U_S \\ Y_S &= C_S\cdot X_S \end{aligned} \tag{3.21}$$

where

$$A_S = \begin{bmatrix} 0 & 1 & 0 & 0 & 0 \\ \frac{2Ci_e^2}{mz_e^3} & 0 & -\frac{2Ci_e^2}{mz_e^3} & 0 & -\frac{2Ci_e}{mz_e^2} \\ 0 & 0 & 0 & 1 & 0 \\ -2k_{q1}\phi_1 C\frac{i_e^2}{z_e^3} & 0 & 2k_{q1}\phi_1 C\frac{i_e^2}{z_e^3} - \omega_1^2 & -2\eta_1\omega_1 & 2k_{q1}\phi_1 C\frac{i_e}{z_e^2} \\ 0 & \frac{i_e}{z_e} & 0 & -\frac{i_e}{z_e} & -\frac{Rz_e}{2C} \end{bmatrix}$$
$$B_S = \begin{bmatrix} 0 & 0 & 0 & 0 & z_e/2C \end{bmatrix}^{\mathrm{T}}, \quad C_S = \begin{bmatrix} 1 & 0 & -1 & 0 & 0 \end{bmatrix} \tag{3.22}$$

Substituting into specific numerical values, the eigenvalue of $A_S$ is calculated to get

$$\lambda_{1,2} = -0.33 \pm 314.52\cdot i,\ \ \lambda_{3,4} = -11.63 \pm 17.14\cdot i,\ \ \lambda_5 = 18.44 \tag{3.23}$$

Therefore, the open-loop system has positive pole. An unstable system requires the design of feedback controller. It can be proved by further analyzing the controllability of the system that the rank of system controllability matrix $\mathrm{rank}\left(\begin{bmatrix} B_S & A_S B_S & {A_S}^2 B_S & {A_S}^3 B_S & {A_S}^4 B_S \end{bmatrix}\right) = 5$ can be obtained. The system's state is completely controllable. Therefore, the pole of closed-loop system can be configured arbitrarily to optimize the controllability as long as the state feedback controller is constituted properly.

However, some problems are found in terms of designing the state feedback controller directly. The first problem is that not all the states are measureable. For example, the state of track $(z_G,\ \ddot{z}_G)$ cannot be measured directly. In addition, in order to ensure reliability of the system, the direct measureable quantity is used for system feedback generally. The second problem is that the direct design is unfavorable to instruct the practical debugging because of the unclear effect of every feedback parameter in the system. As a result, the stable control of the system is finally realized by combining the system features and applying the separate design method approximately.

### 3.3.2 *Particularity of Vehicle–Track Coupling*

In the process of suspension, the electromagnet and the track are coupled with each other by means of the electromagnetic force instead of mechanical contact. As it is expressed in system model, the coupling item is shown in Formula (3.24):

$$z = z_m - z_G \tag{3.24}$$

In Formula (3.24), considering that the track motion may result in the change of gap between the electromagnet and the track, the size of electromagnetic force will change. Therefore, the coupling item Formula (3.24) plays a great influence on coupling between the electromagnet and the track. It is allowed to use the approximately separating design method for the system by means of a certain approximating hypotheses.

The influence of the track motion on the electromagnet motion is analyzed. Suppose that the electromagnet is in a stable suspension state, the electromagnetic force $F_e$ is identically equal to electromagnet's own gravity $mg$ at this moment. The track motion with the action of constant force $F_e$ is analyzed. Based on Formula (3.10), it is solved below:

$$\begin{cases} z_G(t) = \frac{k_{q1}\phi_1 \cdot F_e}{\omega_1^2} \cdot \left(1 - \frac{1}{\sqrt{1-\eta_1^2}} \cdot \mathrm{e}^{-\eta_1\omega_1 \cdot t} \cdot \sin\left(\omega_1\sqrt{1-\eta_1^2} \cdot t + \varphi\right)\right) \\ \varphi = \arctan\frac{\sqrt{1-\eta_1^2}}{\eta_1} \qquad (0 < \eta_1 < 1, \quad 0 < \varphi < \pi/2) \end{cases} \tag{3.25}$$

By combining the actual system parameters, it can be found that the track is a self-stability system with the action of constant electromagnetic force and it is stabilized finally at $z_G(t_e)$. Calculate the equilibrium position of the track.

$$z_G(t_e) = \frac{k_{q1}\phi_1 mg}{\omega_1^2} = 3.04 \times 10^{-5}\ \mathrm{m} \tag{3.26}$$

It can be found that the track motion shows very small amplitude, which may be approximate to small range interference, relative to the electromagnet motion. Suppose the track be ideal steel body firstly and separation hypothesis I is listed as follows:

$$\Delta z_G = 0, \quad \Delta z = \Delta z_m \tag{3.27}$$

At first, the suspension controller ignoring track motion will be designed based on separation hypothesis I.

### 3.3.3 Design of the Suspension Controller Ignoring Track Motion

If the track motion under the Hypothesis Condition (3.27) is ignored, the model obtained by simplification from Formula (3.27) is listed as follows:

$$\begin{cases} m\ddot{z} = mg - C \cdot \frac{i^2}{z^2} \\ u = R \cdot i + \frac{2C}{z} \cdot \dot{i} - \frac{2C \cdot i}{z^2} \cdot \dot{z} \end{cases} \tag{3.28}$$

The equilibrium point of the system can be shown below:

$$\begin{cases} mg = C \cdot \frac{i_e{}^2}{z_e{}^2} \\ u_e = R \cdot i_e \end{cases} \tag{3.29}$$

The equation of electromagnet voltage and current will be analyzed, considering the electromagnet motion within a small range near the equilibrium position. At this moment, the electromagnet inductance may be approximated at a constant:

$$L_e = 2C/z_e \tag{3.30}$$

The equation of electromagnet voltage and current at the equilibrium point is converted as follows:

$$u = R \cdot i + L_e \cdot \dot{i} = R \cdot i + \frac{2C}{z_e} \cdot \dot{i} \tag{3.31}$$

Hence, the voltage and current transfer function of the electromagnet can be obtained:

$$G_i(s) = \frac{I(s)}{U(s)} = \frac{1/R}{\frac{2C}{Rz_e} \cdot s + 1} \tag{3.32}$$

Formula (3.32) indicates that the electromagnet is a first-order inertial link, with time constant $T_i = 2C/Rz_e = 0.206$. As the big delay between the output current $i$ and the control voltage $u$ influences the system design, the current feedback will be introduced firstly [5]. The design index is to decrease the time constant.

The electromagnet within the system's master response frequency band (dozens of hertz) is approximated at one percentage link and the proportionality coefficient is 1.

Set the input voltage as $\hat{u}$ and design the current feedback below:

$$u = k_{c1} \cdot (\widehat{u} - k_{c2} \cdot i) \tag{3.33}$$

Substitute Formula (3.32) into (3.33) and get below the formula

$$\widehat{G}_i(s) = \frac{I(s)}{\widehat{U}(s)} = \frac{k_{c1}/\,(R + k_{c1}k_{c2})}{\frac{2C}{z_e\cdot(R+k_{c1}k_{c2})}\cdot s + 1} \tag{3.34}$$

If the time constant $T_i = 0.0025$ and the proportionality coefficient are set according to the design index, $k_{c1} = 363$ and $k_{c2} = 1$ can be obtained. The current feedback is shown in Fig. 3.2.

Set the input voltage as the step signal and the simulation results before and after the calibrations are shown in Fig. 3.3.

Based on the simulation curves, it can be found that the introduction to current feedback increases the current response speed by a large margin. Therefore, the electromagnet after calibration may be approximated at proportionality link with gain as equal to 1 within the response frequency band of the system:

$$i = \widehat{u}/R_i = \widehat{u}/1 = \widehat{u} \tag{3.35}$$

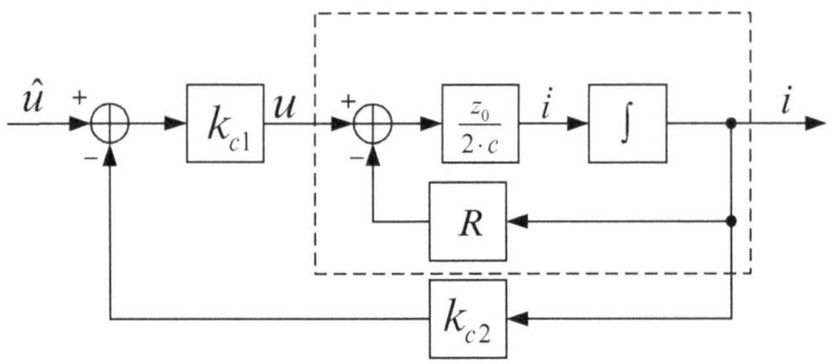


**Fig. 3.2** Current feedback

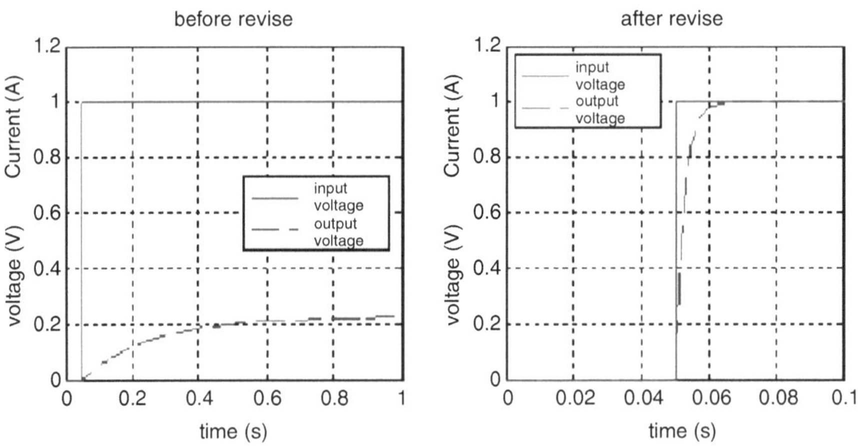


**Fig. 3.3** Simulation curve of current loop before and after calibration

By substituting into modular form of electromagnet Formula (3.28), the original system reduces its order modal to the second-order modal of nonlinear system:

$$m\ddot{z} = mg - C \cdot \frac{\widehat{u}^2}{z^2} \tag{3.36}$$

Use Taylor expansion for linearization of Formula (3.36) at the equilibrium point and the formula below is obtained:

$$\ddot{z} = \frac{2g}{z_e} \cdot z - \frac{2\sqrt{Cg/m}}{z_e} \cdot \widehat{u} \tag{3.37}$$

Suppose that the reference gap is $z_i$ and design the feedback control law as follows:

$$\widehat{u} = \widehat{k}_p \cdot (z - z_i) + \widehat{k}_d \cdot \dot{z} \tag{3.38}$$

Substitute into Formula (3.37) and carry out Laplace Transform to get the transfer function below:

$$G_z(s) = \frac{Z(s)}{Z_i(s)} = \frac{\frac{2\sqrt{Cg/m}}{z_e} \cdot \widehat{k}_p}{s^2 + \frac{2\sqrt{Cg/m}}{z_e} \cdot \widehat{k}_d \cdot s + \frac{2\sqrt{Cg/m}}{z_e} \cdot \widehat{k}_p - \frac{2g}{z_e}} \tag{3.39}$$

Set the performance index to 5 % in overshoot and 0.1 s in regulation time. The feedback parameters can be obtained by comparing the second order of standard model $\widehat{k}_p = 3489$, $\widehat{k}_d = 55.9$. The control law of the electromagnet by combining Formulas (3.23) and (3.38) can be also obtained:

$$u = k_{c1}\widehat{k}_p \cdot (z - z_i) + k_{c1}\widehat{k}_d \cdot \dot{z} - k_{c1}k_{c2} \cdot i \triangleq k_p \cdot (z - z_i) + k_d \cdot \dot{z} - k_c \cdot i \tag{3.40}$$

Since the gap sensor, acceleration sensor, and current sensor are installed in the actual system, the state quantities may be measured directly including suspension gap $z$, acceleration speed of electromagnet $\ddot{z}_m$, and current $i$. Therefore, the speed feedback in Formula (3.40) has been actually obtained by means of the vertical acceleration integral of electromagnet. At the same time, the integral feedback is introduced by static error to eliminate the error. In general, it is deemed that the integral feedback plays a very small influence on dynamic performance of the system and does not change other parameters temporarily. Then the control law is finally obtained as follows:

$$u = k_p \cdot (z - z_i) + k_d \cdot \dot{z}_m + k_I \cdot \int (z - z_i)\mathrm{d}t - k_c \cdot i \tag{3.41}$$

In the formula above, $k_p = 1.267 \times 10^6$, $k_d = 2.029 \times 10^4$, $k_I = 1.815 \times 10^7$, $k_c = 358.6$. Such hypotheses are made in the foregoing analysis that the track is rigid body and the electromagnet is approximated at proportionality link and so on. In order to verify the reasonability of such hypotheses, it is necessary to substitute the designed control law Formula (3.41) into the single-point nonlinear model directly as shown in Formula (3.17), which can realize the simulation verification. Suppose that the initial gap of the electromagnet is 20 mm and the steady state gap is 12 mm. The step response of the system is investigated below [6, 7].

Figure 3.4 shows the change curves of electromagnet displacement, gap, and track displacement, respectively. The electromagnet displacement meets the design requirements basically, and it can respond to the step signal quickly and keep stable at 12 mm. The integral feedback may play a certain influence on dynamic performance of the electromagnet so that deviation occurs between the overshoot and response time of the system and the design index. This deviation may be revised by adjusting other parameters. The hypothesis that the electromagnet is the proportionality link and the design of controller is correct can be found.

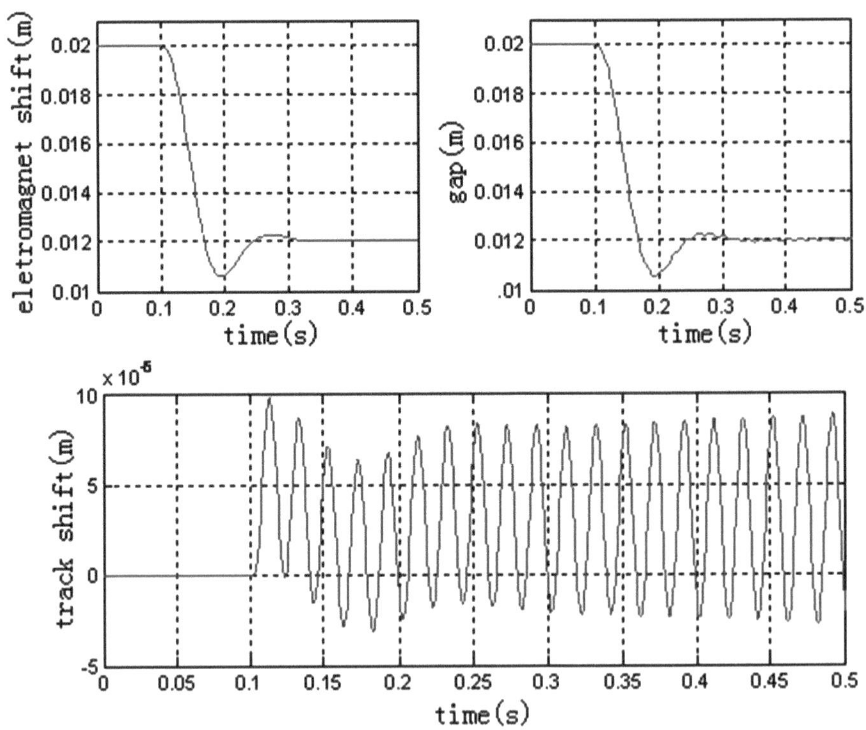


**Fig. 3.4** Step response curve of controller if ignore track motion

The gap curve and the electromagnet displacement curve are roughly the same except for the oscillation within a small amplitude and vibration which cannot converge. For the track displacement, small oscillation with 50 Hz in frequency occurs, and the vibration amplitude shows a tendency to disperse slowly along with time.

From the above, by applying the control law in Formula (3.41), the electromagnet may keep fast trace and set the gap changes and remain stable. However, the continuous vibration happens to the track, and it shows a tendency to disperse gradually, which is well in accordance with the experimental phenomenon. If the control law in Formula (3.41) is applied when the vehicle is in the state of static suspension on the steel beam track, the phenomenon of steady train body with continuous vibration of the track will occur, and the track vibration will disperse slowly and result in the unstable train body finally.

It is very necessary to analyze the vibration reason again. The track itself is a self-stability system that can converge rapidly and keep steady for constant disturbance. The track vibration is the result of continuous motivation of electromagnetic force, and the continuous change of the electromagnetic force arises because the track vibration is fed back into the controller by means of gap quantity. Therefore, it is expected to improve the controller design to get the suspension controller that can suppress the track vibration.

### *3.3.4 Design of Controller Under Relative Static Hypothesis of Electromagnet*

The electromagnet may be basically stabilized in the equilibrium position under the controller effect. The calculation indicates that the electromagnet displacement has $2.41 \times 10^{-6}$ m in maximum fluctuation value within 1 s after the steadiness of system suspension. And it may be ignored approximately relative to the track displacement. Therefore, suppose that the electromagnet is static in the equilibrium position and separation hypothesis II is proposed as follows:

$$\Delta z_m = 0, \quad \Delta z = -\Delta z_G \tag{3.42}$$

Design the controller through ignoring electromagnet motion based on separation hypothesis II. The design index of the controller is able to suppress the vibration rapidly to keep the track stable. The linearization track model separated from Formula (3.43) can be obtained below:

$$\begin{cases} \ddot{z}_G = \left(2k_{q1}\phi_1 C\frac{i_e^2}{z_e^3} - \omega_1^2\right) \cdot z_G - 2\eta_1\omega_1 \cdot \dot{z}_G + 2k_{q1}\phi_1 C\frac{i_e}{z_e^2} \cdot i \\ \dot{i} = -\frac{i_e}{z_e} \cdot \dot{z}_G - \frac{Rz_e}{2C} \cdot i + \frac{z_e}{2C} \cdot u \end{cases} \tag{3.43}$$

As the hypothesis of static electromagnet is valid in the case of applying the controller as shown in Formula (3.41), it is required to substitute into Formula (3.43). If the error integral feedback is ignored, the linearization model of the track closed-loop system can be obtained as follows:

$$\begin{cases} \ddot{z}_G = \left(2k_{q1}\phi_1 C \cdot \frac{i_e^2}{z_e^3} - \omega_1^2\right) \cdot z_G - 2\eta_1\omega_1\dot{z}_G + 2k_{q1}\phi_1 C \cdot \frac{i_e}{z_e^2} \cdot i \\ \dot{i} = -\frac{k_p z_e}{2C} \cdot z_G - \frac{i_e}{z_e} \cdot \dot{z}_G - \frac{k_c z_e + R z_e}{2C} \cdot i \end{cases} \tag{3.44}$$

In order to facilitate the analysis, set

$$\begin{cases} k_1 = 2k_{q1}\phi_1 C \cdot \frac{i_e^2}{z_e^3} - \omega_1^2 = -9.8096 \times 10^4, \ k_2 = -2\eta_1\omega_1 = -0.6280 \\ k_3 = 2k_{q1}\phi_1 C \cdot \frac{i_e}{z_e^2} = 0.4379, \ k_4 = -\frac{k_p z_e}{2C} = -1.3943 \times 10^6 \\ k_5 = -\frac{i_e}{z_e} = -1,141.7, \ k_6 = -\frac{k_c z_e + R z_e}{2C} = -399.68 \end{cases} \tag{3.45}$$

Select the state variable

$$X = \begin{bmatrix} z_G & \dot{z}_G & i \end{bmatrix}^{\mathrm{T}} \tag{3.46}$$

The eigenmatrix of closed-loop system is listed below:

$$A_{S1} = \begin{bmatrix} 0 & 1 & 0 \\ k_1 & k_2 & k_3 \\ k_4 & k_5 & k_6 \end{bmatrix} \tag{3.47}$$

The eigenvalue of $A_{S1}$ is calculated:

$$\lambda_1 = 0.48 + 315.01\mathrm{i}, \ \lambda_2 = 0.48 - 315.01\mathrm{i}, \ \lambda_3 = -401.3 \tag{3.48}$$

As the closed-loop system has a pair of conjugate poles slightly bigger than 0 in real part, the oscillation with approximately 50 Hz in frequency may produce the track displacement $z_G$, and it will disperse slowly along with time, which is in accordance with the abovementioned simulation results.

It is necessary to introduce new feedback quantity to design the controller. The new feedback quantity is expected to contain the information on track motion which is available conveniently at the same time. The track, where the sensor is not installed, cannot be measured directly. According to the above analysis, it can be found that the track vibration may influence the gap measurement, and therefore the gap signal contains the motion information of track in the gap signal and especially under the stable electromagnet, and the gap differential is approximately equal to the speed of track motion after reversed phase. It is also easy to solve the differential under the condition of known gap signal. So the gap differential as new feedback quantity to design the track controller is selected. Hence, the control law of the track can be obtained below:

$$u = k_p \cdot (z - z_i) - k_c \cdot i + k_{dz} \cdot \dot{z} \tag{3.49}$$

Replace Formula (3.41) to get new linear closed-loop model as follows:

$$\begin{cases} \ddot{z}_G = \left(2k_{q1}\phi_1 C \cdot \frac{i_e^2}{z_e^3} - \omega_1^2\right) \cdot z_G - 2\eta_1\omega_1\dot{z}_G + 2k_{q1}\phi_1 C \cdot \frac{i_e}{z_e^2} \cdot i \\ \dot{i} = -\frac{k_p z_e}{2C} \cdot z_G - \left(\frac{i_e}{z_e} + \frac{k_{dz} z_e}{2C}\right) \cdot \dot{z}_G - \frac{k_c z_e + R z_e}{2C} \cdot i \end{cases} \tag{3.50}$$

Compared with Formula (3.47), the closed-loop eigenmatrix of the new system is shown below:

$$A_{S2} = \begin{bmatrix} 0 & 1 & 0 \\ k_1 & k_2 & k_3 \\ k_4 & k_5 - \frac{k_{dz} z_e}{2C} & k_6 \end{bmatrix} \tag{3.51}$$

Use Routh Criterion to analyze the stability of the system and get the condition for stable closed-loop system.

$$k_{dz} > 1287 \tag{3.52}$$

In order to accelerate the convergence speed of track vibration, select $k_{dz} = 1.452 \times 10^4$. The eigenvalue of $A_{S2}$ is calculated below:

$$\lambda_1 = -4.94 + 318.83\text{i}, \;\; \lambda_2 = -4.94 - 318.83\text{i}, \;\; \lambda_3 = -390.42 \tag{3.53}$$

The track vibration will converge rapidly after the introduction of gap differential feedback, and finally the track will keep in stable state. Meanwhile, if control law in Formula (3.41) and control law in Formula (3.49) is taken together, the integrated control law of the system can be obtained:

$$u = k_p \cdot (z - z_i) + k_d \cdot \dot{z}_m + k_{dz} \cdot \dot{z} - k_c \cdot i \tag{3.54}$$

### 3.3.5 *Stability Analysis of Integrated Closed-Loop System and System Simulation*

After introducing the two separation hypotheses, the integrated control law as shown in Formula (3.54) can be obtained, which considers both electromagnet and track motion at the same time by means of step-by-step design. The method of step-by-step design simplifies the design process and defines every feedback variable influences on the control performance. However, it requires further researches on the stability of the integrated control law that is applied in the single-point suspension system. Substituting Formula (3.54) into (3.49), the linearization model of single-point closed-loop system can be obtained below:

$$\begin{cases} \ddot{z}_G = \left(2k_{q1}\phi_1 C\frac{i_e^2}{z_e^3} - \omega_1^2\right)\cdot z_G - 2\eta_1\omega_1\cdot\dot{z}_G - 2k_{q1}\phi_1 C\frac{i_e^2}{z_e^3}\cdot z_m + 2k_{q1}\phi_1 C\frac{i_e}{z_e^2}\cdot i \\ \ddot{z}_m = \frac{2Ci_e^2}{mz_e^3}\cdot z_m - \frac{2Ci_e^2}{mz_e^3}\cdot z_G - \frac{2Ci_e}{mz_e^2}\cdot i \\ \dot{i} = \frac{k_p z_e}{2C}\cdot z_m + \left(\frac{i_e}{z_e} + \frac{k_{dz}z_e + k_d z_e}{2C}\right)\cdot\dot{z}_m \\ \quad -\frac{k_p z_e}{2C}\cdot z_G - \left(\frac{i_e}{z_e} + \frac{k_{dz}z_e}{2C}\right)\cdot\dot{z}_G - \frac{Rz_e + k_c z_e}{2C}\cdot i \end{cases} \tag{3.55}$$

Select the state variable:

$$X = \begin{bmatrix} z_m & \dot{z}_m & z_G & \dot{z}_G & i \end{bmatrix}^{\mathrm{T}} \tag{3.56}$$

Get the eigenmatrix of closed-loop system:

$$A_{S3} = \begin{bmatrix} 0 & 1 & 0 & 0 & 0 \\ 1633 & 0 & -1633 & 0 & -1.43 \\ 0 & 0 & 0 & 1 & 0 \\ -499.9 & 0 & -9.81\times 10^4 & -0.628 & 0.438 \\ 1.39\times 10^6 & 3.95\times 10^4 & -1.39\times 10^6 & -1.71\times 10^4 & -400 \end{bmatrix} \tag{3.57}$$

The eigenvalue of $A_{S3}$ is calculated below:

$$\lambda_{1,2} = -7.9 \pm 319i,\ \ \lambda_{3,4} = -176.8 \pm 104i,\ \ \lambda_5 = -31 \tag{3.58}$$

It can be found that the application of the integrated control law as shown in Formula (3.54) may stabilize the system. It may be deemed that the gap differential is approximately equal to the acceleration integral when the electromagnet moves. It is equivalent to increasing the speed feedback in the electromagnet controller. And for the system as shown in Formula (3.39), it is easy to find that increasing the speed feedback can show the influence on dynamic performance instead of changing the stability of the system, which may be calibrated by adjusting other parameters. Therefore, the integrated control law of the system is obtained:

$$u = k_p\cdot(z - z_i) + k_d\cdot\dot{z}_m + k_{dz}\cdot\dot{z} + k_I\cdot\int(z - z_i)\mathrm{d}t - k_c\cdot i \tag{3.59}$$

In order to achieve the control performance index as originally setting, the control parameters are adjusted by eliminating the influence of error integral feedback and gap differential feedback. The parameters are set as follows:

$$\begin{aligned} k_p &= 1.7424\times 10^6, k_d = 2.029\times 10^4, k_{dz} = 1.452\times 10^4, \\ k_I &= 1.815\times 10^7, k_c = 358.6 \end{aligned} \tag{3.60}$$

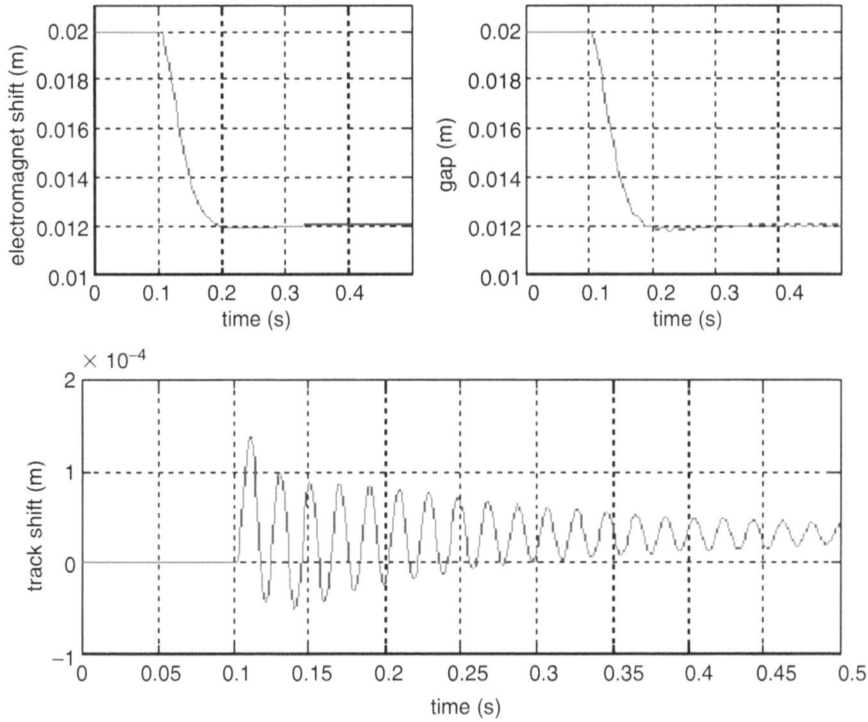


**Fig. 3.5** Step response curves of integrated controller

Use the control law in Formula (3.59) and the adjusted parameters, and substitute into the nonlinear single-point suspension model Formula (3.17) for simulation verification. Suppose that the initial gap of the electromagnet is 20 mm and the steady state gap is 12 mm, the step response of the system will be analyzed below.

Figure 3.5 shows the step response curves of the system after the application of integrated controller. It is obvious that the electromagnet displacement can respond to the step signal quickly and keep stable to meet the dynamic performance index of the electromagnet. The amplitude of the track vibration attenuates rapidly to meet the requirement for track stability. Based on the simulation results, the integrated controller as designed in this section can achieve the expected performance index. It can be found that the proposed method of separate design in this chapter is feasible. In addition, it is found that in the experiment, the integrated controller is effective to suppress the phenomenon of track vibration and ensure the vehicle static to suspend on the steel beam track for a long time or pass the turnouts at a low speed. The results of the integrated controller can improve the control effect of the original system to a large extent and cancel the related restriction on the vehicle operation.

## 3.4 Optimization of Control Parameters Under the Condition of Output Saturation

In particular, it is the default that the controller has "infinite" output capacity in the design process as mentioned above, and the feedback parameters are selected according to the dynamic performance index of the linear closed-loop model to obtain very ideal control effect. However, the output voltage of actual controller, which is restricted by grid voltage of the vehicle, can only change within a certain range. That is equivalent to add a nonlinear saturation link at the output end of the controller. In this section, the research will focus on the design of controller under the conditions that the system has the saturation link.

### *3.4.1 Analysis of Output Saturation Link of the Suspension Controller*

In this chapter, high-power digital chopper is applied at the output end of the suspension controller [8]. The chopper converts the low-voltage control signal to the output of large current and then drives the electromagnet further to achieve the stable suspension of train. The working principle of digital chopper is shown in Fig. 3.6.

In Fig. 3.6, the chopper after simplification is mainly composed of two power triodes and two diodes. Direct current power supply is applied. When the high-level input is at the control end, the triodes are turned on and the current in the electromagnet flows in the direction as shown in Fig. 3.6a. When the low-level input is at the control end, the triodes are cut off. The electromagnet discharges to the

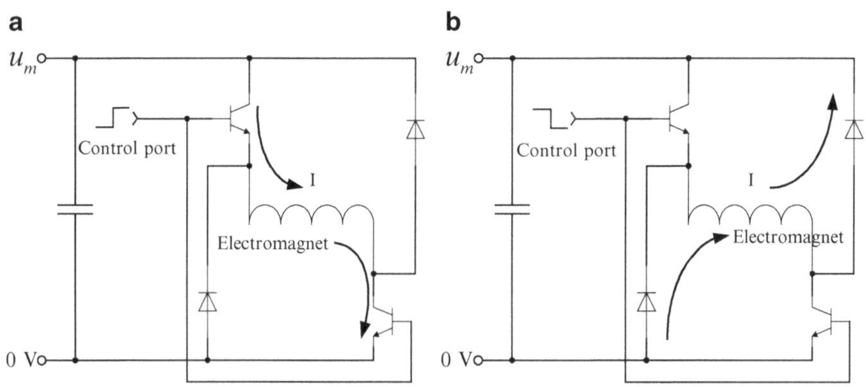


**Fig. 3.6** The working principle of digital chopper. (**a**) Turn-on in event of high level. (**b**) Cutoff in event of low level

capacitor and the grid of the controller through the two diodes, and the current in the electromagnet flows in the direction as shown in Fig. 3.6b.

If the signal that is input to the control end of chopper is the continuous square wave with PWM (pulse-width modulation) 10 kHz width, the control quantity by the control algorithm can manifest the modulation of PWM actual width. The electromagnet may be approximated at an inductance and the current cannot break. Under the effect of high-frequency PWM, the dynamic modulation of PWM width is, at the microlevel, a process continuously charging and discharging rapidly. The control of the control quantity over charging and discharging time proportion, at a macro-level, is approximate to the following output saturation link:

$$\overline{u} = \begin{cases} u_m, & u > u_m \\ u, & -u_m < u < u_m \\ -u_m, & u < -u_m \end{cases} \tag{3.61}$$

In Formula (3.61), $u$ represents the voltage as calculated by controlling algorithm, $u_m$ represents the grid voltage, and $\bar{u}$ represents the voltage as actual operation on both ends of the electromagnet. The grid voltage of train allows a certain range of fluctuation. In order to facilitate the analysis, the lower limit of voltage $u_m = 400$ V is adopted. In particular, Formula (3.61) allows the control quantity to have the negative pressure which actually means that the discharge process of the electromagnet on the grid is equivalent to the controllable negative pressure on both ends of the electromagnet for the low level of PWM.

### 3.4.2 *Design of Optimization of Control Parameters*

For the linearization system as shown in Formula (3.19), the global stability under the effect of control law may be achieved as shown in Formula (3.59). The original closed-loop system is still globally stable after the output saturation link is added as shown in Formula (3.61), or it may lower to the regionally asymptotically stable [9, 10]. For the latter, the size of the stable region is related to the selection of control algorithm and feedback parameters. Suppose that the control algorithm remains unchanged, the method of searching maximum attraction ellipsoid is adopted to realize the design of parameters optimization. Hereby the definitions and related theories are explained in brief below.

For linearization system,

$$\dot{X} = AX + B \cdot \text{sat}(u) \tag{3.62}$$

where sat($u$) represents the saturation function. Considering the state feedback $u = KX$, the corresponding closed-loop system is

$$\dot{X} = AX + B \cdot \text{sat}(KX) \tag{3.63}$$

Suppose that the system (3.63) satisfies the two following conditions. First, the eigenvalues of $(A + BK)$ are all located in the left half plane. Second, $(A,\ B)$ is controllable and Lyapunov equation is guaranteed as follows:

$$(A + BK)^{\mathrm{T}} P + P\,(A + BK) + I = 0 \tag{3.64}$$

There is a sole symmetric positive definite matrix $P$. Thus, one ellipsoid in the state space is determined according to the abovementioned $P$, and a positive number $r$ is as follows:

$$\Omega\,(P, r) = \left\{X \in R^n : X^{\mathrm{T}} PX < r\right\} \tag{3.65}$$

If $X(t) \in \Omega\,(P, r)$, $\forall t > 0$ for the initial state of system (3.63) $\forall X(0) \in \Omega\,(P, r)$, $\Omega(P, r)$ is called as the invariable ellipsoid of system. Define the alternative Lyapunov function $V_P(X) = X^{\mathrm{T}} PX$ of the system (3.62) according to the solution of (3.64). If $\dot{V}_P \leq 0$ for $\forall X \in \Omega\,(P, r)$, $\Omega(P, r)$ is the invariable ellipsoid of system.

Introduce the saturation function as follows:

$$\mu(KX) = \begin{cases} 1\ , & |KX| \leq u_m \\ \frac{u_m}{|KX|}\ , & |KX| > u_m \end{cases} \tag{3.66}$$

Then the saturation function can be represented as $sat(u) = \mu KX$, and the system (3.63) can be represented as follows:

$$\dot{X} = (A + \mu BK)\,X \tag{3.67}$$

It can be found that

$$\begin{aligned} \dot{V}_P(X) &= X^{\mathrm{T}} \left\{(A + \mu BK)^{\mathrm{T}} P + P\,(A + \mu BK)\right\} X \\ &= X^{\mathrm{T}} \left\{\left[(A+BK)^{\mathrm{T}} P + P\,(A+BK)\right] + (\mu - 1) \cdot \left[(BK)^T P + P(BK)\right]\right\} X \\ &= X^{\mathrm{T}} \{-I + (\mu - 1) \cdot E\}\, X \end{aligned} \tag{3.68}$$

where

$$E = (BK)^{\mathrm{T}} P + P(BK) \tag{3.69}$$

It can be proved that when the minimum eigenvalue of $E$ $\lambda_{\min}(E) \geq -1$, $\dot{V}_P \leq 0$ and the output saturation system is still globally stable. When $\lambda_{\min}(E) < -1$, the system is regionally asymptotically stable. The latter's stable region is ellipsoid $\Omega(P, r)$, in which the ellipsoid parameter is shown below:

$$r = \frac{u_m^2}{(KP^{-1}K^{\mathrm{T}})\left(1 + \frac{1}{\lambda_{\min}(E)}\right)^2} \tag{3.70}$$

The volume of ellipsoid is shown below:

$$r = \frac{u_m^2}{(KP^{-1}K^{\mathrm{T}})\left(1 + \frac{1}{\lambda_{\min}(E)}\right)^2} \tag{3.71}$$

where $W$ is the dimensionality of Matrix $A$. Because of the unchanged control law and small change of dynamic performance, it can be considered to apply the aforesaid conclusion to design the suspension controller for solving the maximum stable region by optimizing the feedback parameters.

In the process of designing the controller, it is known that the track displacement is vibration in small amplitude that plays small influence on the control output saturation. The track displacement can be ignored to simplify the analysis. Consider the system model of output saturation link after simplification; it is shown as follows:

$$\dot{X}_L = (A_L + \mu B_L K_L)\, X_L \tag{3.72}$$

The definition of $\mu$ has been shown in (3.66), and the other variables have been defined as follows:

$$\begin{aligned} X_L &= \left[ z_m \; \dot{z}_m \; i \right]^{\mathrm{T}} \\ A_L &= \begin{bmatrix} 0 & 1 & 0 \\ \frac{2Ci_e^2}{mz_e^3} & 0 & -\frac{2Ci_e}{mz_e^2} \\ 0 & \frac{i_e}{z_e} & -\frac{Rz_e}{2C} \end{bmatrix} \\ B_L &= \left[ 0 \; 0 \; z_e/2C \right]^{\mathrm{T}}, \quad K_L = \left[ k_p \; k_d + k_{dz} \; -k_c \right] \end{aligned} \tag{3.73}$$

In the feedback parameters, $k_{dz}$ plays the great influence on suppression force of track vibration, and $k_c$ is the key parameter for reducing the delay of the electromagnet. Due to the fact that there has almost no space for optimization, the influence of $k_p$ and $k_d$ on stable region will be mainly analyzed. In order to reduce the impact of parameter optimization on system dynamic performance, the numerical value as shown in Formula (3.60) is taken as the initial value of parameters to change $k_p$ and $k_d$ bidirectionally with $\alpha$ and $\beta$:

$$\begin{aligned} k_p(i) &= k_p(0) + (i - n/2) \cdot \alpha, \quad i = 0, 1, \ldots, n \\ k_d(j) &= k_d(0) + (j - n/2) \cdot \beta, \quad j = 0, 1, \ldots, n \end{aligned} \tag{3.74}$$

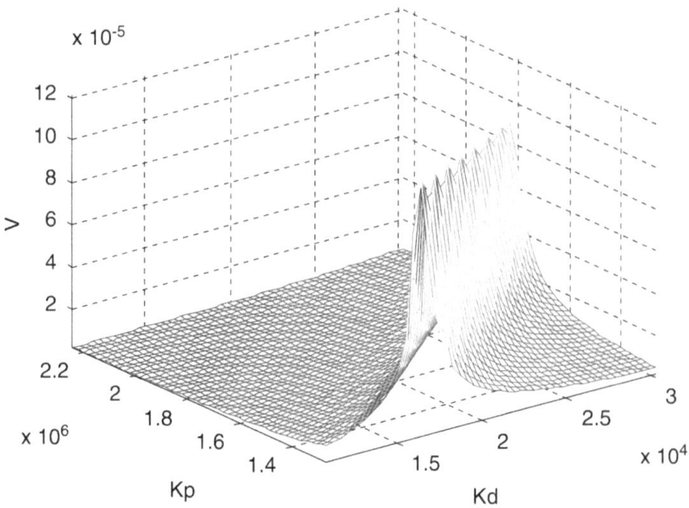


**Fig. 3.7** Changing of curved surface of ellipsoid volume along with parameters

Hence, a state feedback gain set $K_L(i,j)$ can be obtained. The corresponding $P(i,j)$ in regard to Formula (3.64) is calculated. Then, $E(i,j)$ in regard to Formula (3.69) is calculated. If $\lambda_{\min}(E(i,j)) \geq -1$ exists, $k_p(i)$ and $k_d(j)$ are the optimal parameters that can keep the system globally stable. Otherwise, the ellipsoid volume $V(i,j)$ is further combined with Formula (3.70) and (3.71) is calculated. The values $i$ and $j$ that maximize the volume are solved. $k_p(i)$ and $k_d(j)$ are the optimal parameters that can enable the system to have the biggest stable region.

Compare the features of parameters and select $n = 100$, $\alpha = 10{,}000$, $\beta = 200$; the calculation results indicate that $\lambda_{\min}(E(i,j)) < -1$ within the range of parameter change. That is to say, there is no parameter set to keep the saturation system globally stable. Figure 3.7 shows the change of curved surface of the ellipsoid volume along with parameters.

In Fig. 3.7, it can be seen that if $k_p$ keeps unchanged and $k_d$ changes, one value of maximizing the ellipsoid volume $k_d(j_{i,m})$ exists and $k_d(j_{i,m})$ increases gradually with the increase of $k_p$. If $k_d$ keeps unchanged and $k_p$ changes, one value of maximizing the ellipsoid volume $k_p(i_{j,m})$ exists and $k_p(i_{j,m})$ increases gradually with the increase of $k_d$. If $k_p$ and $k_d$ increase in the meantime, the ellipsoid volume decreases accordingly. The state feedback gain of maximizing the ellipsoid volume is calculated as follows:

$$\begin{aligned} &k_p = 1.2824 \times 10^6, k_d = 1.669 \times 10^4, \\ &k_{dz} = 1.452 \times 10^4, k_I = 1.815 \times 10^7, k_c = 358.6 \end{aligned} \quad (3.75)$$

The effect of parameter optimization will be described by means of simulation below.

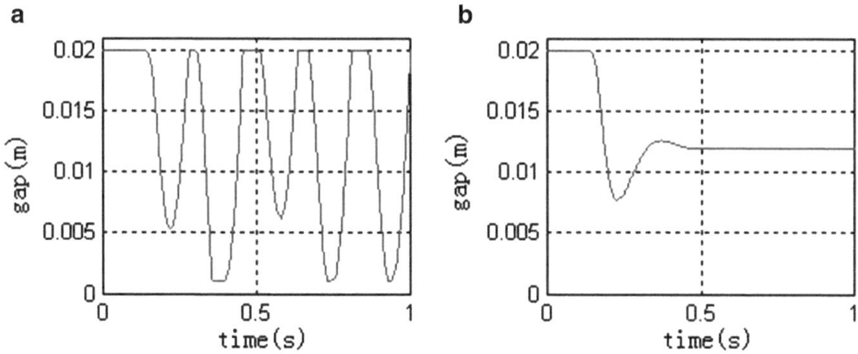


**Fig. 3.8** Gap response curve involving output saturation link. (**a**) Use the original parameters. (**b**) Use the optimized parameters

### *3.4.3 System Simulation Considering the Output Saturation Link*

After the saturation link is added as shown in Formula (3.61) at the output end of the controller, it is substituted into the nonlinear single-point suspension model as shown in Formula (3.17) for simulation verification. Suppose that the initial gap of the electromagnet is 20 mm and the steady-state gap is 12 mm, the controller that adopts the original parameter and the step response in the process of optimizing parameters are compared as shown in Fig. 3.8.

From the gap change curve as shown in Fig. 3.8, it can be found that after the addition of output saturation link to the controller, the system cannot remain stable only by using the original parameters, which can keep stable by using the optimized parameters. Compared with the closed-loop system that adopts the original control parameters, the control system that adopts the optimized parameters has bigger stable region. Compared with Figs. 3.8 and 3.5, it can be found that the parameter optimization can reduce the system dynamic performance which is mainly manifested as increasing overshoot and extending adjustment time. But it is closer to the actual situation of the system. The analysis can point out that the rigidness and damping of system cannot be increased arbitrarily in the process of testing and debugging by considering the system dynamic performance, only as it is possible to result in the "disastrous" consequence.

To sum up, the influence of controller's output saturation on the suspension system mainly adds a certain constraint to the parameter selection. It requires the rigidness and damping of system not to be too big. The analysis of saturation link has a big guiding significance in practice.

In addition, it can be pointed that after the method of separate design is applied in design of controller, the globalization is not so strong in the theoretical level than in overall design. It can highlight the influence of every parameter on the suspension control effect and meanwhile offer one kind of step-by-step debugging. In the practical process, it is easy to realize the vehicle static suspension according to the step-by-step debugging thinking of "current feedback–electromagnet–track whole" and give full play to the guiding role of simulating calculation in practice.

## 3.5 Design of Signal Filter of Suspension Control System

In the process of aforesaid controller design and system simulation, it is supposed that all the feedback quantity is ideal and some problems are ignored, such as noise interference and insufficient measurable quantity in actual system. Thus, it is necessary to combine the features of sensor signal in Maglev system to analyze the problems in terms of the feedback signal by suspension control. In addition, through the systematical analysis of the influence of some problems, such as signal noise, false, and erroneous signals, the specific index requirements for processing the related signals to ensure the desirable suspension control effect are proposed [11].

The sensor configuration in the single-point suspension control system is generally shown in Fig. 3.9. The suspension sensor is installed at the end of the electromagnet. Inside the sensor, gap ($Z$) and acceleration ($A$) are integrated. The gap sensor is used to measure the mechanical gap between the upper surface of electromagnet and the lower surface of track. The acceleration sensor is used to measure the vertical motion acceleration of electromagnet relative to the inertial system.

Voltage ($U_I$, $U_{II}$), current ($I_I$, $I_{II}$) and temperature ($T$) sensors are installed inside the suspension controller. The voltage sensor is used to measure the input and output voltage of the controller. The current sensor is used to measure the input current and output current of the controller. The temperature sensor is used to monitor the temperature inside the controller.

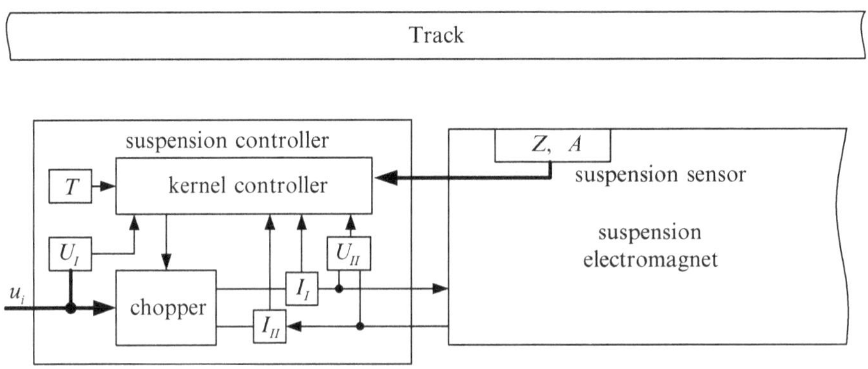


**Fig. 3.9** Configuration diagram of sensor in single-point suspension control system

The kernel controller is used to make comprehensive treatment of the information on all sensors and drive the chopper by generating the control quantity according to the control algorithm. The chopper is used to complete the power current output and drive the electromagnet to produce the electromagnetic force. And the control algorithm ensures that the electromagnetic force is equal to the carrying capacity under the set gap to keep the single point stably suspending.

The signals that are measurable by the single-point system include gap $z$, acceleration $a$, current $i$, input voltage $u_{\mathrm{i}}$, output voltage $u_{\mathrm{o}}$, and temperature $t_p$.

Full-simulation implementation method is applied to the suspension controller in early stage due to the constraint of electronic components. The signal processing and control law are realized by means of the analog network. With the development of high-speed processor DSP, the control law gradually evolves into the digitalized implementation with the retained analog filter network. Thus, the whole suspension control system is still one part of semi-digital implementation way. The suspension system has many sensors with high real-time requirement, and the serial processing mode of DSP cannot meet the requirement for multi-signal filtering. The super-large-scale programmable device FPGA appears and enables the parallel computing of digital filter to be possible [12, 13]. The controller, which is designed in the way of combining FPGA with DSP, may realize the full-digitalization transformation. Compared with the analog controller and semi-digitalized controller, the full-digital controller becomes easier to regulate, has stable parameters, and has the ability to realize the complicated algorithm. It is certain to be the irresistible trend that the full-digital suspension controller will replace full-analog and semi-digital controller. But the previous experimental researches, which have been mainly made on the platform of semi-digitalized controller, have discussed more about digital control law and analog filter and involved less research on digital filter of the suspension system. Consequently, it is required firstly to solve the digitalization implementation of the filter to truly implement the full digitalization of the controller and focus on the researches how to apply digital filtering algorithm to implement filtering functions such as gap low-pass, gap differential and acceleration blocking integral [14].

### 3.5.1 Design of Gap Signal Filter

It is necessary to use the differential signal of gap for the design of control law. The differential value of the gap signal is indirectly obtained by mainly designing the differentiator. Although the gap signal may suppress the part of high-frequency noise through the low-pass filter, it is impossible to eliminate the influence of noise totally. If the common differentiator is used, the high-frequency noise will be enlarged and even effective signal is probably submerged. As a result, it is necessary to design the differential filter which can suppress the high-frequency noise [15, 16]. In the suspension control system, the main role of gap differential feedback is to suppress the phenomenon of vehicle-coupled-track vibration occurring on the steel beam track and the turnouts [17]. Therefore, the differentiator is required to be capable to reflect the role of gap differential better within the frequency band of track vibration. Through the actual measurement, the vibration frequency of steel beam track and the turnouts is approximately 50 Hz. Thus, the design index of the gap differentiator can be proposed, which can play the role to be approximate to the differential in signals with lower than 50 Hz in frequency and no less than 80° in phase advance. It can suppress high-frequency noise at the same time. Hence, it is called as the low-pass differentiator.

As the control cycle is far longer than the data communication cycle and many gap differential signals can be obtained within one control cycle, it is easy to find that the method of mean value can suppress high-frequency noise effectively. Therefore, the direct digital design method is applied to combine data with difference to enable the filter to be approximate to the ideal differential at low-frequency band and show bigger attenuation at high-frequency band. So, the integral coefficient smoothing differential filter considering the features of hardware programming language is designed.

The concept of relative frequency is often applied in the design of digital filter. Suppose that the actual signal frequency is $f$ and the sampling frequency is $f_s$. Define the relative frequency as $f' = f/f_s$ and the circular frequency as $\omega = 2\pi f' = 2\pi f/f_s$.

At first, the export process of optimal low-order low-pass differential filter is described. The integral coefficient smoothing differential filter is the suboptimal differential filter, which is obtained by a certain simplification on the former basis.

The ideal low-pass differential filter should be provided based on the amplitude frequency characteristic of differential in low-frequency band and cutoff in high-frequency band:

$$H_d\left(\mathrm{e}^{j\omega}\right) = \begin{cases} j\omega & |\omega| \le \alpha\pi \\ 0 & \text{other} \end{cases} \tag{3.76}$$

where $\alpha\pi$ means the cutoff frequency and $0 < \alpha < 1$.

It is possible to require the higher order to implement the low-pass differential filter approximating to Formula (3.76) in characteristic by using a general method for the digital filter design. It is considered to use a low-order filter to be approximate to ideal low-pass differentiator and keep the square error between the two smallest. Hence, it is called as the optimal low-order low-pass differential filter that means the optimal differentiator for short.

With an odd function $H(\mathrm{e}^{j\omega})$ passing the origin, it can be approximate to $H_d(\mathrm{e}^{j\omega})$.

$$H\left(\mathrm{e}^{j\omega}\right) = j\sum_{k=1}^{M} C_k \sin\left(k\omega\right) = j\sum_{k=1}^{M} C_k \frac{\mathrm{e}^{jk\omega} - \mathrm{e}^{-jk\omega}}{2} \tag{3.77}$$

The corresponding filter form is shown below:

$$y(n) = \sum_{k=1}^{M} C_k \frac{x(n+k) - x(n-k)}{2} \tag{3.78}$$

where $C_k$ means the coefficient of filter.

Different differentiators, which can be obtained by using different $C_k$, may show different approximations to $H_d(e^{j\omega})$. In order to investigate the approximation, define the error function as follows:

$$E(\alpha, C_k) = \int_{-\pi}^{\pi} \left| H_d\left(e^{j\omega}\right) - H\left(e^{j\omega}\right) \right|^2 d\omega \tag{3.79}$$

Hence, Formula (3.80) can be obtained as follows:

$$E(\alpha, C_k) = \frac{2}{3}(\alpha\pi)^3 + \pi \sum_{k=1}^{M} C_k^2 + 4 \sum_{k=1}^{M} \frac{C_k}{k^2} \left[ k\alpha\pi \cos(k\alpha\pi) - \sin(k\alpha\pi) \right] \tag{3.80}$$

The size of $E(\alpha, C_k)$ is affected by two kinds of parameters, cutoff frequency $\alpha\pi$, and coefficient $C_k$. For the cutoff frequency given, the smaller the $E(\alpha, C_k)$ is, the more degree the $H(e^{j\omega})$ approximates to $H_d(e^{j\omega})$.

Now suppose that $C_k,\ k = 1, 2, \ldots, M$ is known and set

$$\frac{\partial E(\alpha, C_k)}{\partial(\alpha\pi)} = 0 \tag{3.81}$$

Solve the optimum band-pass frequency $\alpha\pi$ by minimizing $E(\alpha, C_k)$, and Formula (3.82) is obtained as follows:

$$\sum_{k=1}^{M} C_k \sin(k\alpha\pi) = \frac{\alpha\pi}{2} \tag{3.82}$$

Namely,

$$H\left(e^{j\alpha\pi}\right) = j\frac{\alpha\pi}{2} \tag{3.83}$$

Formula (3.83) indicates that for given $C_k$, the point of intersection of curve $H(e^{j\omega})$ and straight line $j\frac{\omega}{2}$ (i.e., $\omega = \alpha\pi$) is the optimum band pass of the differentiator. Then suppose $\alpha\pi$ be known and set

$$\frac{\partial E(\alpha, C_k)}{\partial C_k} = 0,\ k = 1, 2, \ldots, M \tag{3.84}$$

Use Formula (3.83) at the same time and there are $M+1$ equations totally. So there are $M+1$ unknown numbers $C_1, C_2, \ldots, C_M$ and $\alpha$ can be solved. These parameters can ensure minimum $E(\alpha, C_k)$ within $0 \sim \alpha\pi$. The method of least square is possible to result in big error at local level in spite of the minimum total error within given interval. In order to make $H(e^{j\omega})$ to be better approximate to $H_d(e^{j\omega})$ in low-frequency band, another constraint condition may be added for the process of solving the equation that is set below:

$$\left.\frac{\mathrm{d}H\left(e^{j\omega}\right)}{\mathrm{d}\omega}\right|_{\omega=0} = j \tag{3.85}$$

Thus, it is equivalent to Formula (3.83) as follows:

$$\sum_{k=1}^{M} kC_k = 1 \tag{3.86}$$

In this way, it may ensure the minimum total error of $H(e^{j\omega})$ within $0 \sim \alpha\pi$ and the approximation to be the ideal difference at low-frequency band. Generally, the numerical analysis is applied to solve the equation set. For example, when $M=3$, Formula (3.83) can be obtained below:

$$C_1 = 0.097, C_2 = 0.1666, C_3 = 0.1897, \alpha = 0.24, E\left(\alpha, C_k\right) = 0.126 \tag{3.87}$$

By substituting the coefficient as shown in Formulas (3.87) into (3.78), the optimal differentiator is obtained for $M=3$. The parameters of optimal differentiator, which are generally not integer, are not convenient to calculate. So a certain simplification of the parameters to get the smoothing differential filter can be considered.

Suppose that the current moment is $n$ and the smoothing differential filter is centered on $n \pm N$. The difference after averaging of $2L+1$ data is calculated according to the following algorithm:

$$y(n) = \frac{1}{2N}\left\{\frac{1}{2L+1}\cdot\sum_{k=-L}^{L} x\left(n+N+k\right) - \frac{1}{2L+1}\cdot\sum_{k=-L}^{L} x\left(n-N+k\right)\right\} \tag{3.88}$$

Select $N = 2,\ L = 1$ according to the sampling cycle and the control cycle of Maglev system; hence, the gap smoothing differential filter can be obtained as shown below:

$$y(n) = \frac{1}{12}\cdot\{[z\,(n+1) - z\,(n-3)] + [z\,(n+2) - z\,(n-2)] + [z\,(n+3) - z\,(n-1)]\} \tag{3.89}$$

It is equivalent to

$$y(n) = \frac{1}{12} \cdot \{[z(n+1) - z(n-1)] + [z(n+2) - z(n-2)] + [z(n+3) - z(n-3)]\} \tag{3.90}$$

The corresponding frequency response can be obtained as follows:

$$H\left(e^{j\omega}\right) = \frac{j}{6} \cdot \{\sin(\omega) + \sin(2\omega) + \sin(3\omega)\} \tag{3.91}$$

Comparing Formulas (3.90) and (3.78), the smoothing differential filter maintains the structure of the optimal differentiator except that parameter $C_k$ is selected differently. The optimum band-pass frequency $\alpha\pi$ and the error $E(\alpha, C_k)$ of the smoothing differential filter to analyze the performance of filter can be calculated below.

The optimum band pass of the smoothing differential filter is calculated to be $\alpha = 0.251$ and the error is $E(\alpha, C_k) = 0.142$. Compared with Formula (3.87), it can be known that the error of the smoothing differential filter is slightly bigger than that of the optimal differentiator and the optimum band pass of both is close. Although the smoothing differential filter loses the accuracy, the complicated decimal computation can be avoided. Thus, the filter can be called as suboptimal low-order low-pass differential filter. Figure 3.10 shows the amplitude frequency response of filter.

Based on Fig. 3.10, it can be found that the smoothing differential filter has approximately ideal difference at low-frequency band and big attenuation at high-frequency band. Its optimum band-pass frequency 5.02 kHz can meet the design requirements. If the filter's gain can be adjusted to enable the signal gain of the

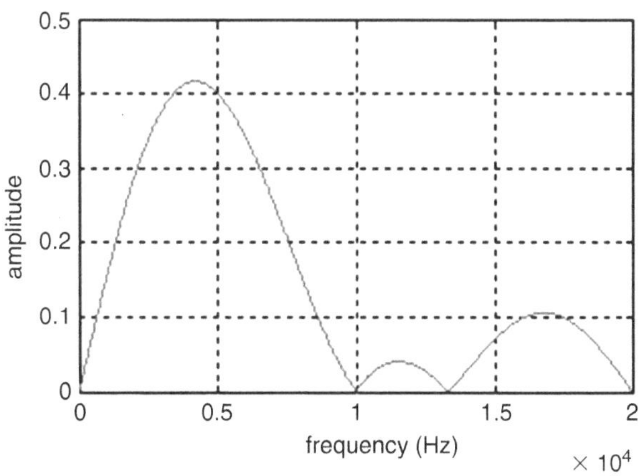


**Fig. 3.10** Amplitude frequency curve of smoothing differential filter

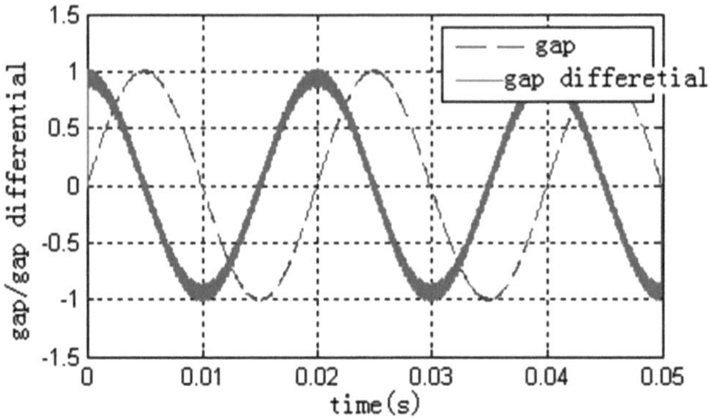


**Fig. 3.11** Simulation curve of smoothing differential filter

filter for 50 Hz to be approximate to 1, the gap smoothing differential filter after adjustment can be obtained as follows:

$$y(n) = 10 \cdot \{[z(n+1) - z(n-1)] + [z(n+2) - z(n-2)] + [z(n+3) - z(n-3)]\} \tag{3.92}$$

Suppose the input be 50 Hz sinusoidal signal that contains the high-frequency interference with 9 kHz in frequency and 0.01 in amplitude. The result of filter simulation is shown in Fig. 3.11.

According to the simulation result in Fig. 3.11, it can be found that the smoothing differential filter can meet the design requirements for phase and amplitude.

### 3.5.2 *Design of Acceleration Signal Filter*

The accelerometer is installed horizontally on the suspension sensor to measure the acceleration of vertical motion of the electromagnet. This installation way can enable the acceleration signal to contain gravity acceleration component so that the speed signal cannot be integrated directly and it is required to design a filter to filter out the gravity disturbance. But the influence of gravity acceleration on measurement is not ideal direct current quantity because suspension control is aimed at the electromagnet that keeps trace of the track with a certain gap in the process of vehicle motion as well as rolling motion of the electromagnet. And, it may arise when the vehicle runs on the nonlinear sections including curves and ramps as contained in the track. The gravity component in acceleration signal will change along with the various gesture of electromagnet. The acceleration signal and the power spectrum as measured in the process of actual train running are shown in Fig. 3.12.

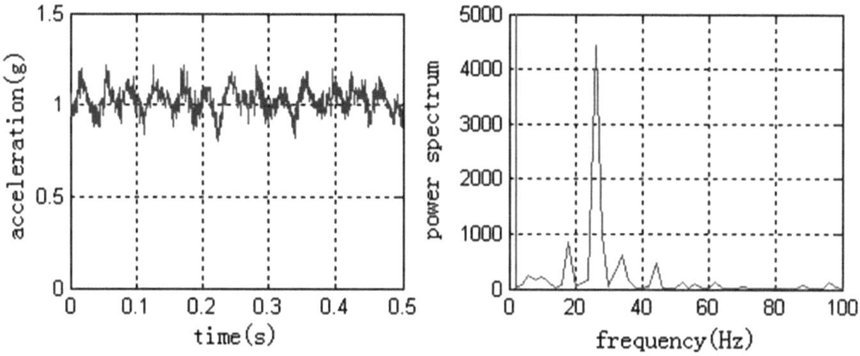


**Fig. 3.12** Analysis of acceleration signal and power spectrum

Based on Fig. 3.12, it can be found that the frequency of acceleration signal is concentrated on the range lower than 50 Hz which mainly contains direct current component and signal around 25 Hz. Figure 3.12 is only a piece of acceleration measurement data in the process of a certain time of running. The acceleration signal will have different performances with different vehicle speeds, running section, and control algorithm. It can be concluded to have the following features according to the observation of lots of experiments as follows:

1. The acceleration is low-frequency signal.

The frequency of acceleration signal is lower than 100 Hz. The main response frequency is 20–60 Hz when the vehicle keeps stationary or runs at a low speed. And about a hundred hertz of signal may occur for some reasons, such as the vibration of electromagnet.

2. Too high sampling frequency.

The frequency of acceleration signal is lower than 100 Hz. Generally, the selected sampling frequency in the process of data acquisition is 3–5 times as much as the highest frequency of signal. Too high sampling frequency will bring big high-frequency interference. But the accelerometer and other signals use the same one sampling module in the design of suspension controller. The sampling frequency is up to 40 kHz. Such frequency is unfavorable to design of digital filter. Therefore, it is required to re-extract the sampling signal.

3. Big gravity disturbance.

The accelerometer, which is directly installed on the electromagnet, is very sensitive to the gesture change of electromagnet. The train shows very complicated running gestures in the line with many influence factors such as irregular track, curves, and ramps. The gravity disturbance changes along with gesture of the electromagnet, especially when it changes greatly in the process of entering and

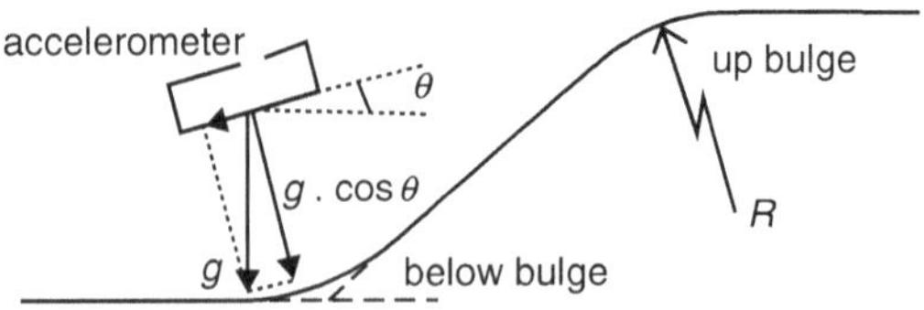


**Fig. 3.13** Electromagnet passes the vertical and circular curve

leaving the ramps. The below part will focus on analyzing the influence of gravity on the acceleration signal when the vehicle enters and leaves the ramps.

The ramp section of the Maglev train track is composed of straight line segments, transition curve, and vertical and circular curve. The transition curve, which is set between the straight line segment and the vertical and circular curve, may improve the comfort in the running process. The electromagnet shows the biggest change amplitude of pitching gesture in the process of passing the vertical and circular curve. The measured gravity fluctuation by accelerometer is the most drastic. Figure 3.13 shows the gravity component that is measured by the accelerometer when the electromagnet passes the vertical and circular curve.

The acceleration as measured is listed below:

$$a(n) = a_v(n) + a_g(n) + \delta(n) \tag{3.93}$$

where $a_v(n)$ represents the electromagnet motion acceleration at $n$ moment, $a_g(n) = g \cdot \cos\theta$ represents the quantity of gravity disturbance, and $\delta(n)$ represents noise interference. $\delta(n)$, which arises mainly from the electromagnetic interference of the device, is approximate to white noise. $\theta$ represents the angle at which the pitching gesture changes when the electromagnet passes the vertical and circular curve, whose angular rate is related to the curve radius and passing speed:

$$\theta = \omega \cdot (nt_s) = \frac{v}{R} \cdot (nt_s) \tag{3.94}$$

where $v$ represents the driving speed, $R$ represents the radius of the vertical curve, and $t_s$ represents the sampling cycle. The radius of the vertical curve is determined according to the requirements for train speed, comfort, and safety. When the angular rate $\omega$ is small, the gravity component may be approximated to a very-low-frequency direct current signal. In order to verify the filtering effect, the analysis with the change quantity of the maximum angular rate (0.043 rad/s) is considered as follows:

$$a_g(n) = g \cdot \cos(0.043 \cdot nt_s) \tag{3.95}$$

The filter that can remove the gravity disturbance effectively according to the features of the acceleration signal is designed. It is called as blocking filter in terms of its features of gravity component approximating to direct current signal. In processing the acceleration signal, the gravity component is required to filter out as the signal of very low frequency. The dynamic component of vertical motion of electromagnet is required to retain. The most direct design idea is to extract the dynamic component by using high-pass filter. However, this method is not good for subsequent processing of signals or design of low noise system because of its amplification effect on high-frequency noise.

According to the features of each component in measured acceleration values as shown in Formula (3.92), an analysis is made below. $a_g(n)$ represents the signal of very low frequency, and $a_v(n)$ represents the intermediate frequency signal. The suspension control is aimed at enabling the electromagnet to keep in a stable state, i.e., $a_v \approx 0$. When big deviation occurs in $a_v$, the controller will suppress the tendency of continuing to enlarge and control to approach zero. If the white noise form of $\delta(n)$ is added, it can be deemed to have an appropriate $N$. Thus,

$$\frac{1}{N}\sum_{k=0}^{N-1}\left[a_v(n-k)+\delta(n-k)\right]\approx 0 \tag{3.96}$$

Meanwhile meets

$$\frac{1}{N}\sum_{k=0}^{N-1}\left[a_g(n-k)\right]\approx a_g(n) \tag{3.97}$$

Get the blocking filter of acceleration signal as follows:

$$\begin{aligned} y(n) &= a(n)-\frac{1}{N}\sum_{k=0}^{N-1}\left[a(n-k)\right] \\ &= a(n)-\frac{1}{N}\sum_{k=0}^{N-1}\left[a_g(n-k)+a_v(n-k)+\delta(n-k)\right] \\ &\approx a(n)-a_g(n)=a_v(n)+\delta(n) \end{aligned} \tag{3.98}$$

Focus on analyzing how to select $N$. The transfer function of $N$ mean filter is shown below:

$$H(z)=\frac{1}{N}\sum_{k=0}^{N-1}z^{-k}=\frac{1}{N}\cdot\frac{1-z^{-N}}{1-z^{-1}} \tag{3.99}$$

The obtained frequency response is shown below:

$$H\left(e^{j\omega}\right)=\frac{1}{N}\cdot e^{-j\omega(N-1)/2}\cdot\frac{\sin(\omega N/2)}{\sin(\omega/2)} \tag{3.100}$$

The amplitude response of filter is one *sinc* function, and the amplitude at $\frac{2\pi}{N}k,\ k=0,1,\ldots,N-1$ is zero. The unilateral bandwidth of its main lobe is $2\pi/N$. Figure 3.14 is the amplitude frequency response diagram of mean filter when $N=9$.

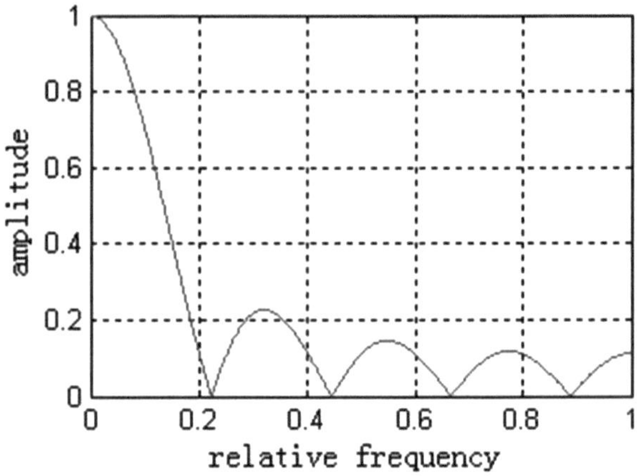


**Fig. 3.14** Amplitude frequency response diagram of mean filter

It can be found that mean filter is a low-pass filter. The bandwidth of main lobe based on the features of acceleration signal to enable the filter to meet the requirements is selected as shown in Formula (3.97).

Based on Formula (3.99) and the foregoing analysis of the acceleration signal, it can be found that the bandwidth of main lobe should be bigger than the maximum frequency of $a_g(n)$ change and smaller than the lowest frequency of $a_v(n)$ main frequency band. Select 2.5 Hz for the main lobe bandwidth by combining the measured data. It is known that the sampling frequency is 40 kHz. Hence, it is obtained that the acceleration filter parameter is $N = 4 \times 10^4/2.5 = 1.6 \times 10^4$.

The obtained $N$ value is too big and it is very difficult to design such filter in actual condition. The reason for too big $N$ value is mainly too high sampling frequency of the acceleration signal. So it is required to re-extract the acceleration signal. Select 320 Hz for re-extracting frequency; hence, the filter parameter after re-extracting $N = 320/2.5 = 128$ can be obtained.

Set a FIFO array with 128 in length $x(k),\ k = 0, 1, \ldots, 127$ when the algorithm is implemented. Extract the acceleration sampling data every other 124 ones to put them in the FIFO array and calculate the gravity component as follows:

$$\widehat{a}_g = \frac{1}{128}\sum_{k=0}^{127} x(k) \tag{3.101}$$

Before the re-extracting of signals, the original data require the first-order low-pass anti-aliasing filtering to prevent the data after re-extracting from aliasing. Then subtract the gravity component from the original signal. Hence, the acceleration signal after blocking can be obtained below:

$$\tilde{a}(n) = a(n) - \widehat{a}_g \tag{3.102}$$

The acceleration blocking filter is applied to the measured waveform as shown in Fig. 3.12. The curve after filtering is shown in Fig. 3.15.

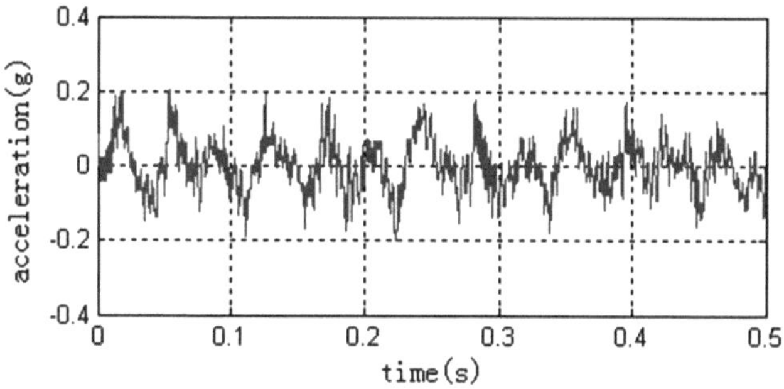


**Fig. 3.15** Output curve of blocking filter

It can be found from the figure that the blocking filter can effectively filter out the gravity disturbance and meanwhile truly reflect the acceleration of electromagnet motion. The blocking filter is easy for realization owing to its simple algorithm.

## 3.5.3 *The Delay Influence of Gap Low-Pass Filter on Suspension Control*

In order to eliminate the high-frequency noise, the gap signal uses the first-order low-pass filter. For convenience of analysis, it is shown as follows:

$$G_{LP}(s) = \frac{1}{T \cdot s + 1} \tag{3.103}$$

where $T > 0$ represents the time constant of low-pass filter. Big $T$ value will gain better effect of low-pass filtering and meanwhile result in the bigger signal delay. The time constant shows its influence on stability of closed-loop system in three aspects. First, it narrows the selection scope of control parameters. Second, it changes the distribution of poles in the original system. Third, it influences the characteristic of gap differential filter.

### 3.5.3.1 Constraint of Time Constant on Selection of Control Parameters

Ignore the delay of gap differential signal and take no account of elastic track. Focus on analyzing how the delay of gap signal affects the controller of electromagnet. Retain the original current feedback and substitute Formula (3.102) into the control law as shown in Formula (3.38) and obtain the control law including gap low-pass filter as follows:

$$\widehat{u} = \widehat{k}_p \cdot \left( \frac{z}{T \cdot s + 1} - z_i \right) + \widehat{k}_d \cdot \dot{z} \tag{3.104}$$

Substitute into Formula (3.37); hence, the simplification model of the electromagnet including gap low-pass filter is obtained as follows:

$$G_{ZLP}(s) = \frac{\frac{2\sqrt{Cg/m}}{z_e}\widehat{k}_p(Ts+1)}{Ts^3 + \left(\frac{2\sqrt{Cg/m}}{z_e}\widehat{k}_d T + 1\right)s^2 + \left(\frac{2\sqrt{Cg/m}}{z_e}\widehat{k}_d - \frac{2g}{z_e}T\right)s + \frac{2\sqrt{Cg/m}}{z_e}\widehat{k}_p - \frac{2g}{z_e}} \tag{3.105}$$

Use Routh Criterion of Stability to analyze the scope of parameter $\widehat{k}_p$ and $\widehat{k}_d$ to get the following set of inequalities:

$$\begin{cases} \widehat{k}_d > \sqrt{\frac{mg}{C}} \cdot T \\ \left(\frac{2\sqrt{Cg/m}}{z_e}\widehat{k}_d^2 - \frac{2g}{z_e}\widehat{k}_d \cdot T + \frac{\widehat{k}_d}{T}\right) > \widehat{k}_p > \sqrt{\frac{mg}{C}} \end{cases} \tag{3.106}$$

If the low-pass filter is not considered, the parameters of original system $\widehat{k}_p$ and $\widehat{k}_d$ can be selected within the scope as shown below:

$$\begin{cases} \widehat{k}_d > 0 \\ \widehat{k}_p > \sqrt{\frac{mg}{C}} \end{cases} \tag{3.107}$$

Compare Formula (3.105) with Formula (3.106); it can be seen that the delay of gap signal narrows the selection scope of control parameters. Compared with the actual system parameters, the time constant $T$ is not more than 0.01 s generally. Substitute into Formula (3.105); hence obtain

$$\widehat{k}_d > 11.42 \tag{3.108}$$

According to the analysis in Sect. 3.3, it can be known that $\widehat{k}_d$ can generally conform to Formula (3.107). Get the relation of $\widehat{k}_p$, and the value range, with $T$ and $\widehat{k}_d$ according to Formula (3.105), is shown in Fig. 3.16.

As is shown in Fig. 3.16a, the upper limit of $\widehat{k}_p$ is related to $T$ and $\widehat{k}_d$. If $T$ is fixed, the upper limit of $\widehat{k}_p$ decreases with the decrease of $\widehat{k}_d$. If $\widehat{k}_d$ is fixed, the upper limit of $\widehat{k}_p$ decreases rapidly with the increase of $T$. Figure 3.16b shows the curve of relation between upper limit of $\widehat{k}_p$ and $T$ if $\widehat{k}_d = 50$.

So it can be found that the delay of gap low-pass filter can change the selection scope of original parameters and especially shows great influence on gap feedback parameter $\widehat{k}_p$. When the time constant of the filter increases, the value range of $\widehat{k}_p$ will decrease rapidly with it, and the system is not easy to keep stable.

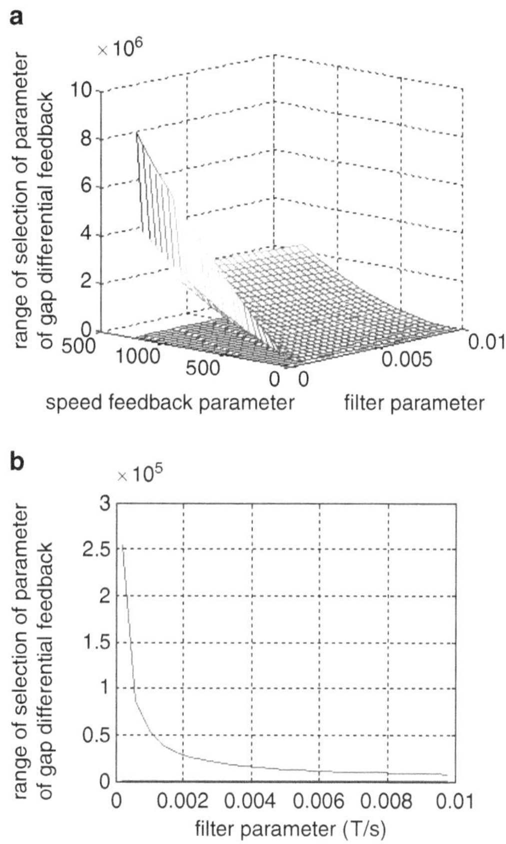


**Fig. 3.16** Relation between selection of gap feedback parameter and delay of signal. (**a**) Graph of relation with speed feedback. (**b**) Graph of relation with $T$ if $\widehat{k}_d = 50$

#### 3.5.3.2 Influence of Time Constant on Poles in the Closed-Loop System

Suppose the parameters of original system $\widehat{k}_p$ and $\widehat{k}_d$ keep unchanged. The system after introduction of gap low-pass filter is shown in Formula (3.104). Compared with the original closed-loop system (3.39), it can be found that the usage of filter adds one closed-loop null point and one closed-loop pole and meanwhile changes the distribution of two poles in the original closed-loop system.

In order to analyze how the time constant affects the null points and the poles of the closed-loop system, suppose that $T$ changes within 0.01 s, hence the curve of changes of null points and poles in the closed-loop system (3.104) along with time constant $T$ is obtained as shown in Fig. 3.17.

It can be found from Fig. 3.17 that after introduction of gap low-pass filter, if $T$ changes within 0.01 s, the pole and null point as added to the system are far from the imaginary axis (exceeding the display range of Fig. 3.17). Therefore, their influence on the system can be ignored. But the time constant $T$ will change the distribution of the two leading poles in the original system, and the two leading poles gradually

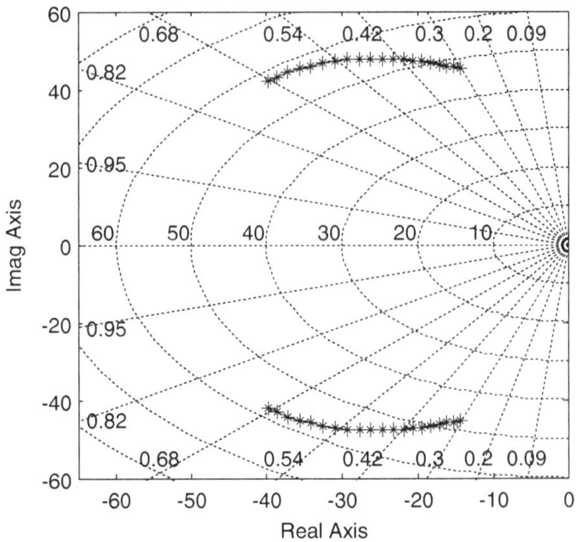


**Fig. 3.17** Distribution of null points and poles in closed-loop system of the electromagnet

approach the imaginary axis with the increase of $T$ value so that the damping ratio of the system decreases along with it.

Therefore, the introduction of gap low-pass filter changes the distribution of the leading poles in the original system under the condition of unchanged feedback parameters. When $T$ increases, the damping ratio of the system will decrease gradually. The below part will explain further about how the change of time constant affects the system dynamic performance by means of simulation.

#### 3.5.3.3 Influence of Time Constant on Gap Differential Filter

The input signal of gap differential filter comes from the output of gap low-pass filter. Therefore, the gap differential signal will produce a certain delay likewise. For the convenience of analysis, the gap differential can be represented as follows:

$$\dot{z} = \frac{s}{T \cdot s + 1} \cdot z \tag{3.109}$$

where $T > 0$ represents the time constant of low-pass filter. Gap differential feedback has the main role of suppressing track vibration. The time constant influences the system in two major aspects. First, it narrows the parameter selection scope of gap differential feedback. Second, it weakens the system's ability to suppress the track vibration. They will be discussed respectively below:

1. Constraint on selection of control parameters

Ignore the delay of gap signal and suppose that the electromagnet keeps in stable state. Analyze how the delay of gap differential signal influences the track motion.

Substitute Formula (3.79) into track control law as shown in (3.49), and the track control law including low-pass filter is obtained as follows:

$$u = k_p \cdot (z - z_i) - k_c \cdot i + k_{dz} \cdot \frac{s \cdot z}{T \cdot s + 1} \tag{3.110}$$

Substitute into Formula (3.43) and the track closed-loop model including low-pass filter is obtained as follows:

$$\begin{cases} \ddot{z}_G = \left(2k_{q1}\phi_1 C \cdot \frac{i_e^2}{z_e^3} - \omega_1^2\right) \cdot z_G - 2\eta_1\omega_1\dot{z}_G + 2k_{q1}\phi_1 C \cdot \frac{i_e}{z_e^2} \cdot i \\ \ddot{i} = \left(-\frac{i_e}{z_e} \cdot \left(2k_{q1}\phi_1 C \cdot \frac{i_e^2}{z_e^3} - \omega_1^2\right) - \frac{k_p z_e}{2C} \cdot \frac{1}{T}\right) \cdot z_G \\ \quad + \left(\frac{i_e}{z_e} \cdot 2\eta_1\omega_1 - \frac{k_p z_e}{2C} - \frac{i_e}{z_e} \cdot \frac{1}{T} - \frac{z_e}{2C} \cdot \frac{k_{dz}}{T}\right) \cdot \dot{z}_G \\ \quad + \left(-\frac{i_e}{z_e} \cdot 2k_{q1}\phi_1 C \cdot \frac{i_e}{z_e^2} - \frac{k_c z_e + R z_e}{2C} \cdot \frac{1}{T}\right) \cdot i + \left(-\frac{k_c z_e + R z_e}{2C} - \frac{1}{T}\right) \cdot \dot{i} \end{cases} \tag{3.111}$$

Select the state variable:

$$X = \begin{bmatrix} z_G & \dot{z}_G & i & \dot{i} \end{bmatrix}^{\mathrm{T}} \tag{3.112}$$

Still use the symbol definition as shown in Formula (3.45) and get eigenmatrix of closed-loop system as follows:

$$A_{LP} = \begin{bmatrix} 0 & 1 & 0 & 0 \\ k_1 & k_2 & k_3 & 0 \\ 0 & 0 & 0 & 1 \\ k_1 k_5 + k_4/T & k_2 k_5 + k_4 + \frac{k_5 - z_e k_{dz}/(2C)}{T} & k_3 k_5 + k_6/T & k_6 - 1/T \end{bmatrix} \tag{3.113}$$

Substitute into specific numerical values and get the system's eigen equation through the calculation below:

$$\begin{aligned} |\lambda I - A_{LP}| = {} & \lambda^4 + \left(401 + \frac{1}{T}\right) \cdot \lambda^3 + \left(9.90 \times 10^4 + \frac{401}{T}\right) \cdot \lambda^2 \\ & + \left(4.01 \times 10^7 + \frac{0.483 \cdot k_{dz} + 9.90 \times 10^4}{T}\right) \cdot \lambda \\ & + \left(1.73 \times 10^{-9} + \frac{4.01 \times 10^7}{T}\right) \end{aligned} \tag{3.114}$$

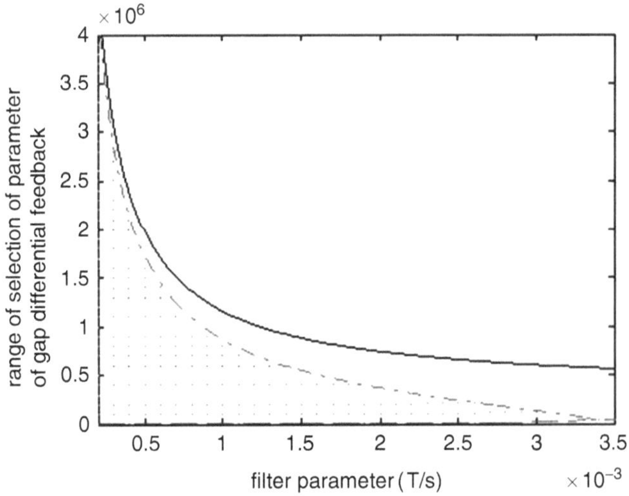


**Fig. 3.18** Relation between parameter selection of gap differential feedback and signal delay

Use Routh Criterion of Stability to analyze the scope of parameter $k_{dz}$ to get the following set of inequalities:

$$\begin{cases} k_{dz} < \dfrac{830}{T} - 8.30 \times 10^5 \cdot T + 3.33 \times 10^5 \\ k_{dz}^2 + \left(8.40 \times 10^7 \cdot T - \dfrac{831}{T} - 1.28 \times 10^5\right) \cdot k_{dz} + 6.90 \times 10^{13} \cdot T^2 \\ \quad + 1.70 \times 10^{11} \cdot T + 6.90 \times 10^8 + \dfrac{1.72 \times 10^6}{T} < 0 \end{cases} \tag{3.115}$$

According to Formula (3.114), it can be obtained that when $T$ value changes. The value range of $k_{dz}$ is shown in the shaded part of Fig. 3.18.

From Fig. 3.18, it can be found that when the time constant $T$ increases, the value range of $k_{dz}$ decreases rapidly. When $T > 0.0034$ s, there is no way to select suitable $k_{dz}$ to stabilize the system. Compared with Formula (3.52), it can be found that the delay of gap differential signal changes the selection scope of parameter $k_{dz}$. When the time constant of filter increases, the value range of $k_{dz}$ will decrease rapidly, and the system is not easy to keep stable.

2. Influence on poles in the system

Suppose that the original system parameters are kept unchanged, and the system's eigenmatrix after the introduction of gap low-pass filter is shown in Formula (3.112). Compared with the eigenmatrix of the original system (3.51), it can be found that after the introduction of gap low-pass filter, the system adds one closed-loop pole and meanwhile changes the distribution of poles in the original system.

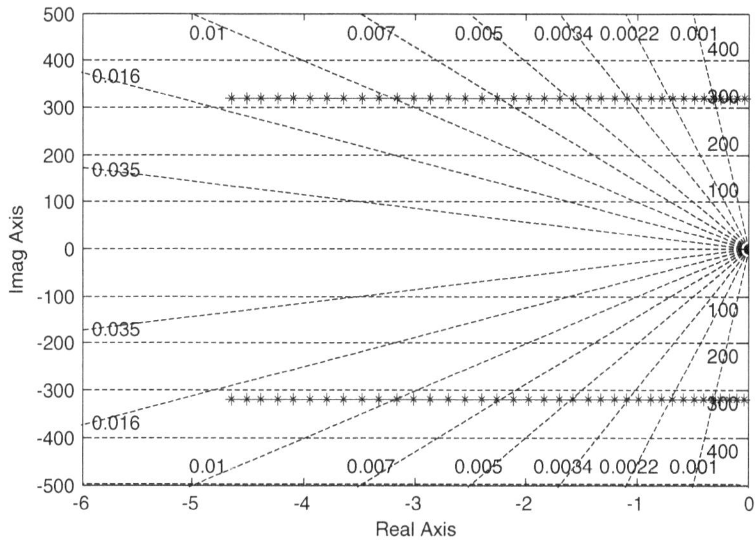


**Fig. 3.19** Distribution of poles in the track closed-loop system

In order to analyze how the time constant influences the poles in the closed-loop system, suppose $T$ change within 0.0034 s. The curve of changes of poles in the system along with time constant $T$ is obtained as shown in Fig. 3.19.

From Fig. 3.19, it can be found that after introduction of gap low-pass filter, the pole as added to the system and the non-leading poles in the original system are far from the imaginary axis (exceeding the display range of the figure). Therefore, their influence on the control system can be ignored. But the time constant $T$ changes the distribution of the two leading poles in the original system, and the two leading poles gradually approach the imaginary axis with the increase of $T$ value. Considering the decrease of system's damping ratio along with it, the track vibration will converge in longer duration.

### 3.5.3.4 Selection Basis of Time Constant

The gap signal and gap differential signal delay have been analyzed independently mainly for the purpose of their different influence on the system. It can be found that the gap signal and gap differential signal delay have some aspects in common in terms of their influence on the system. First, the signal delay increases the constraint on selection of feedback parameters; in particular, the bigger the delay is, the smaller the value range of the parameters is. Second, the signal delay decreases the system's damping ratio; in particular, the bigger the delay is, the smaller the system's damping is. Meanwhile, there are some differences between the both, which is manifested mainly in more sensitivity of gap differential signal to the delay and higher requirement for selection of upper limit of the time constant $T$.

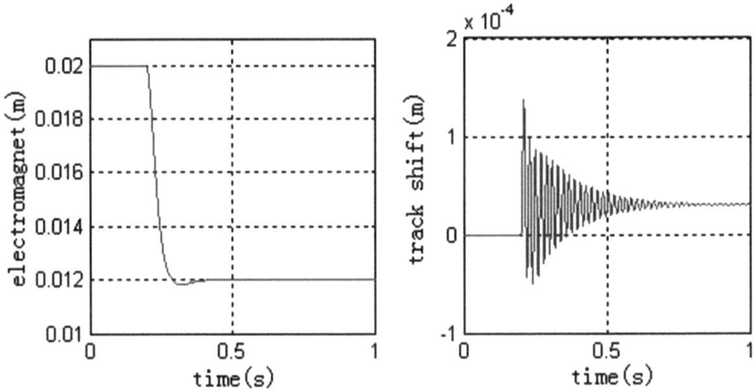


**Fig. 3.20** Curve of step response in event of $T = 0.0002$ s

As a result, in the process of selecting $T$ value, it is mainly considered to satisfy the restriction of gap differential signal on delay. $T = 0.0002$ s is generally selected so that the system's ability to suppress track vibration would not be influenced. Thus, the integrated control law of the system that considers both gap signal and gap differential signal delay can be obtained as follows:

$$u = k_p \cdot \left( \frac{z}{T \cdot s + 1} - z_i \right) + k_d \cdot \dot{z}_m + k_{dz} \frac{s}{T \cdot s + 1} \cdot z + k_I \cdot \int \left( \frac{z}{T \cdot s + 1} - z_i \right) \mathrm{d}t - k_c \cdot i \quad (3.116)$$

Substitute into nonlinear single-point suspension model as shown in (3.17) for simulation verification, and the step response curve of the system can be obtained as shown in Fig. 3.20.

Compared with Figs. 3.20 and 3.5, it can be found that when the time constant is $T = 0.0002$ s, the gap signal and gap differential signal delay of low-pass filter show almost no influence on the dynamic performance of the system and meet the design requirement. In this section, the influence of gap low-pass filter on the system is mainly analyzed and the filter parameter selection is proposed. For the performance analysis of other filters, it is required to add the filter to the system. In addition, the experimental verification is required to be completed for all the filters in the actual system at last.

## 3.6 Experimental Research

After the theoretical analysis and simulation verification, a set of suspension control systems including digital signal filter system and control algorithm can be tested. The system availability can be finally verified upon the actual suspension test which mainly involves the following major test items:

1. Heaving suspension test

Heaving suspension process refers to the vehicle moving from descending state to stable suspension state, namely, the process of the electromagnet gap changing from the initial maximum value (20 mm or so) to preset suspension gap (10 mm). The slow heaving algorithm is added to enhance the comfort, which can enable the gap to change slowly to the preset gap and keep stable.

2. Descending test

Descending process refers to the vehicle moving from instable suspension state to descending state, namely, the process of the electromagnet gap changing from the initial preset suspension gap (10 mm) to the initial maximum value (20 mm or so). The slow descending algorithm is added as well to enhance the comfort, which can enable the vehicle to descend slowly on the track.

3. Static suspension test

Static suspension means that the vehicle keeps in stable suspension state and has no horizontal motion relative to the track.

Observing the filter output signal in the test items above may help to verify the correctness of filter design. In order to make it compatible with the semi-digitalized suspension controller, the digital output of signal processing circuit is transformed to the analog signal output by D/A convertor. In order to eliminate the D/A transformation ripple, D/A output signal is required to go through a first-order low-pass filter before being transmitted to the control part. The related test signals are acquired by oscilloscope from the output end, and the acquired data are redrawn by means of MATLAB.

### *3.6.1 Signal Test of Gap Channel*

The test signal of gap channel includes the gap and gap differential signals. The signal curves change, respectively, in the process of heaving suspension, descending, and static suspension. The gap and gap differential curves in the heaving suspension process are shown in Fig. 3.21.

The gap and gap differential curves in the descending process are shown in Fig. 3.22.

When the control algorithm does not use the feedback information of gap differential signal, the phenomenon of vehicle-coupled-track vibration will occur in static suspension. The fluctuation curve of gap is shown in Fig. 3.23.

When the control algorithm introduces gap differential feedback, it can effectively suppress the track vibration. The gap and gap differential curves of static suspension are shown in Fig. 3.24.

It is worth pointing out that there is some occasional peak interference in the curves due to the fact that the electromagnetic interference increases on the

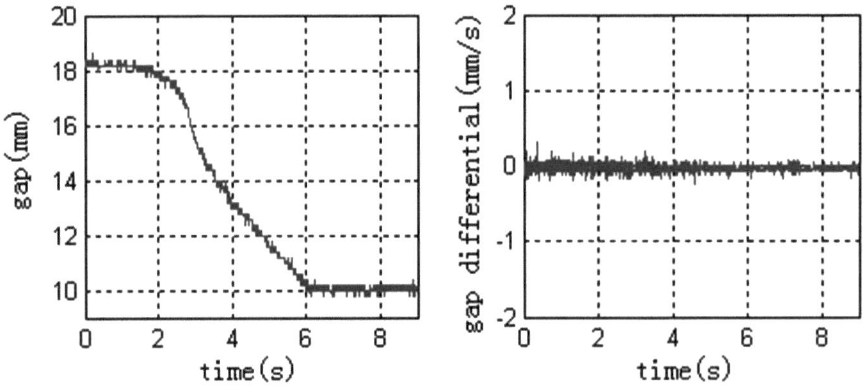


**Fig. 3.21** Gap and gap differential curves in the heaving suspension process

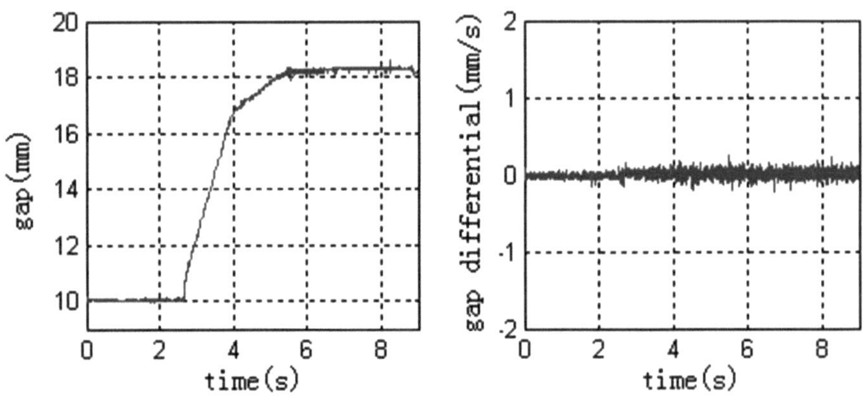


**Fig. 3.22** Gap and gap differential curves in the descending process

oscilloscope and has nothing with the filter output (the same as below). Based on the curves of Figs. 3.21, 3.22, 3.23, and 3.24, it can be seen that the heaving suspension and descending process of vehicle are slow and steady for ensuring comfortable running. Considering the slow heaving suspension and descending process of vehicle, the gap differential signal is approximate to a straight line, which indicates that the differentiator can suppress the noise interference in the system effectively.

If the gap differential feedback is not used in static suspension, the phenomenon of vehicle-coupled-track vibration will be easy to occur and the fluctuation frequency of the gap curve in the equilibrium position is about 60 Hz. At the moment, the gap differential is the fluctuation signal with the same frequency, the phase advance is about 90°, and the alternating current gain is about 1, which can meet the design requirements. After the introduction of gap differential feedback, the control system can effectively suppress track vibration to guarantee stable suspension of the vehicle, and the gap can keep almost unchanged, which indicates that the vehicle has high suspension stability.

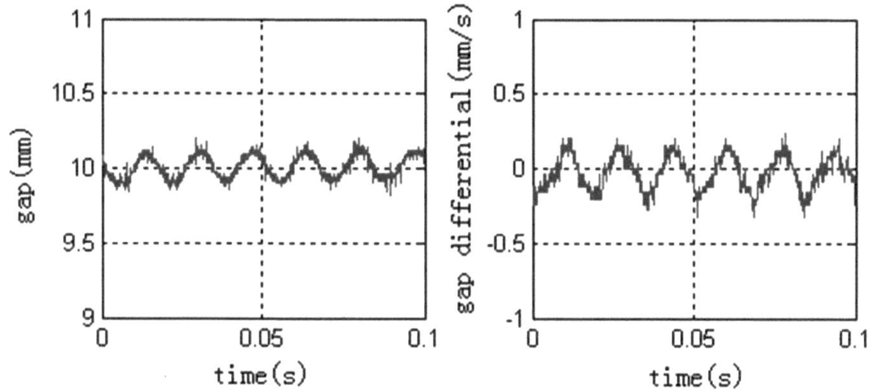


**Fig. 3.23** Gap and gap differential curves in case of vibration

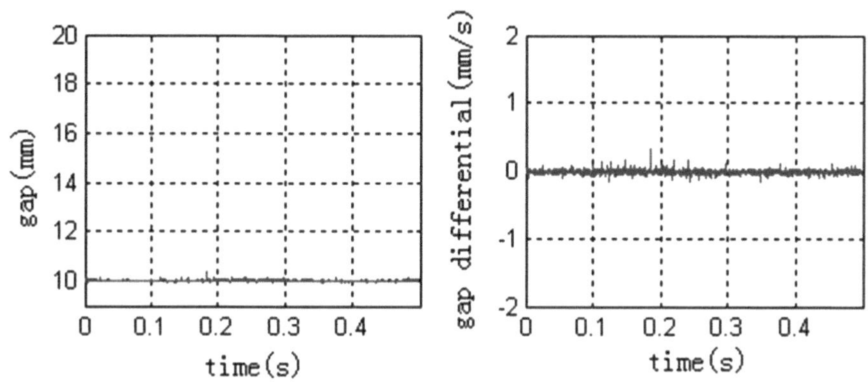


**Fig. 3.24** Gap and gap differential curves of static suspension

### *3.6.2 Signal Test of Acceleration Channel*

The test signal of acceleration channel contains the acceleration signal and acceleration integral signal. The curve of static suspension is shown in Fig. 3.25.

For the shock of big mass in a test point, the acceleration and acceleration integral curves of the suspension point are shown in Fig. 3.26.

From Figs. 3.25 and 3.26, it can be seen that the acceleration integral is a fluctuating curve for static suspension, which indicates that the acceleration blocking filter can effectively eliminate the gravity disturbance. For the shock of suspension point, the various curves of the acceleration integral indicate that the blocking filter does not influence the dynamic change of the acceleration and can meet the design requirements. The tests above have proven that the vehicle has high stability for static suspension and the designed filter can be successfully applied to the actual system and obtain satisfactory experimental effect.

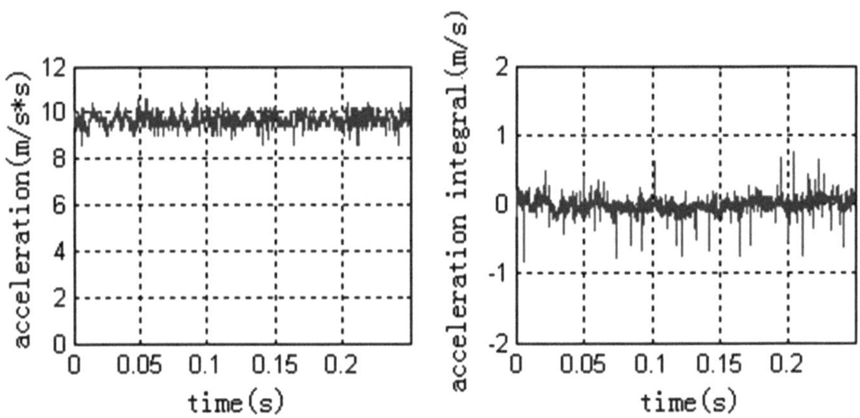


**Fig. 3.25** Acceleration and acceleration integral curves in case of static suspension

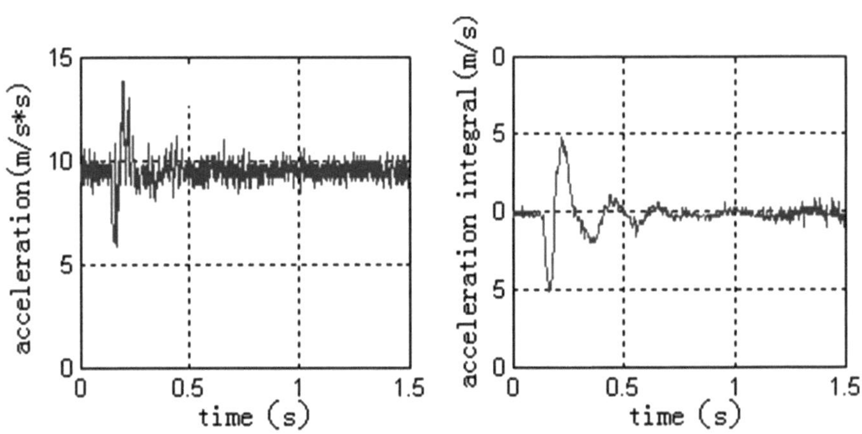


**Fig. 3.26** Acceleration and acceleration integral curves for outage of adjacent points

## 3.7 Conclusion

First, the single-point suspension model is built and the reason of track vibration is analyzed. Then, the suspension controllers are respectively designed by combining the separate design with simulation. By introducing the gap differential feedback, the control algorithm is implemented, which can effectively suppress the track vibration under the specific condition. Finally, the output saturation limitation of the suspension system is considered, and the results of parameter optimization under the condition of output saturation are given. The simulation results indicate that the proposed method of making separate design for controller in this chapter is feasible.

The corresponding digital filter is designed by combining the characteristics and control demands of different sensors in the suspension control system. The implementation of filter is considered, and the delay influence of gap filter for the performance of whole suspension control system is analyzed. Finally, it can be verified that the filter performance can meet the demands of system by means of system simulation and engineering tests.

## References


1. Sinha PK. Electromagnetically suspension dynamics control [M]. London: Peter Peregrinus Ltd; 1987.
2. Liu Hengkun, Chang Wensen, Shi Xiaohong. Vehicle-coupled-guideway vibration in Maglev system [J]. Comput Simul. 2006;23(9):256–258. 267.
3. Katsuhiko Ogata. Modern control engineering [M]. 4th ed. Beijing: Publishing House of Electronics Industry; 2003.
4. Hu Shousong. Auto control elements [M]. Beijing: National Defense Industry Press; 1994.
5. Liu Hengkun, Chang Wensen. Double-loop control of Maglev train [J]. Control Eng China. 2007;14(2):198–200.
6. Chen Guiming, Zhang Mingzhao. Using MATLAB modeling and simulation [M]. Beijing: Science Press; 2001.
7. Chen Yayong. Detailed MATLAB signal processing [M]. Beijing: Posts & Telecom Press; 2001.
8. Wang Ning. Study on Maglev chopper of low and middle speed EMS Maglev [D]. PhD thesis. Changsha: University of Defense Technology; 2004.
9. Wang Jinhua. Analysis for linear systems with actuator saturation [D]. PhD thesis. Qingdao: Qingdao University; 2007.
10. Wei Airong. Saturation control system stability and interference suppression [D]. PhD thesis. Jinan: Shandong University; 2006.
11. Hu Guangshu. Digital signal processing principles, algorithms, applications [M]. Beijing: Tsinghua University Press; 2003.
12. Armstrong JR, Gail Gray F. VHDL design representation and synthesis [M]. Beijing: China Machine Press; 2002.
13. Wang Cheng, Wu Jihua, Fan Lizhen. Altera FPGA/CPLD design [M]. Beijing: Posts & Telecom Press; 2005.
14. Proakis JG, Manolakis DG. Digital signal processing: principles, algorithms, and applications [M], vol. 4. London: Pearson Education Group; 2006.
15. Wei Wei. MATLAB toolbox control engineering technical manual [M]. Beijing: National Defense Industry Press; 2004.
16. Xu Xin, Li Tao, Bo Xiaochen. Matlab toolbox guide-control engineering [M]. Beijing: Electronic Industry Press; 2000.
17. Long Zhiqiang, Hao Aiming, Chang Wensen. Suspension controller design of Maglev train considering the rail track periodical irregularity [J]. J Univ Def Technol. 2003;25(2):84–9.

# Chapter 4
# Control and Diagnosis System of Maglev Train

## 4.1 Introduction

The control and diagnosis system of Maglev train includes onboard control system and diagnosis system [1–3]. The train control system is mainly used to obtain the status signals to monitor and control the onboard equipment, which includes two major functions [4, 5]. The first function is to process the status signals produced by the onboard equipment and then transmit them to the onboard control system. The second function is to assign the control orders given by onboard control system to the related equipment after processing [6]. The onboard control system is mainly consisted of the onboard central processing unit, digital I/O module, analog I/O module, control switch, information process channel, control equipment object, corresponding display module, and so on.

## 4.2 Onboard Control and Diagnosis System

### *4.2.1 Functional Requirements and System Composition*

The control and diagnosis systems can ensure the safety and stable operation of Maglev train, and the major functions are listed as follows:

1. Control and monitor the operation of Maglev train to ensure the proper manipulation for the driver. Master the operation status of Maglev train in real time to ensure safe, quick, and comfortable operation of the train.
2. Monitor the onboard equipment at all levels in Maglev train to ensure they work coordinately under the manipulating instructions of the driver and driving commands.



Z. Liu et al., *Maglev Trains*, Springer Tracts in Mechanical Engineering,
DOI 10.1007/978-3-662-45673-6_4

3. Require the real-time information detection for train control and diagnosis, including the process data, the message data, and the monitoring data. Monitor and identify the faults in the operation process exactly and accurately. Prompt the driver to take corresponding measures to eliminate the faults in time.
4. Record and store the information of equipment faults in the operation process such as time, location, current values of the related parameters, and their changes. These data are required to be shown on the display screen and sent out by the train-ground wireless data transmission device to the ground system for further analysis and processing, which can provide information and data for the train's overhauling, reduce the maintenance duration, and save the maintenance cost.
5. Save and show the record of the operation condition, operation time, important environmental status data, etc., which are completed by driver in the operation process of train.
6. Indicate the driver about the operation mode in terms of train faults, including giving the proposals of maintenance function and mode, which can help to eliminate faults rapidly.

In general, the control and diagnosis system of Maglev train is requested to monitor train's overall operation and acquire, analyze, process, show, and store information such as the status and faults of control equipment and controlled equipment that are distributed in different vehicles [7, 8].

The system is composed of the following major parts [9].

#### 4.2.1.1 Basic Structure

The structure of onboard control and diagnosis system is shown in Fig. 4.1. The system is mainly composed of the following parts:

- *Onboard controller*

  Namely, CCU/CCUR is used to complete the functions such as information monitoring and processing, generation of control instructions, and data storage.
- *Controlled object*

  They are low-voltage cabinet, high-voltage cabinet, DC110V control power supply, DC330V suspension power supply, AC380V auxiliary inverter, suspension controller, traction inverter, brake control system, door unit, air condition, and other equipments.
- *Control switch*

  Driver's keys, control handle, and control switch, which are used for the driver's manipulation.
- *Display unit*

  They are status monitoring and fault diagnosis assessment computer, operation control signal screen, indicator lights, and simulation meter, which are used to display data such as the status and fault diagnosis of both the train and equipment.

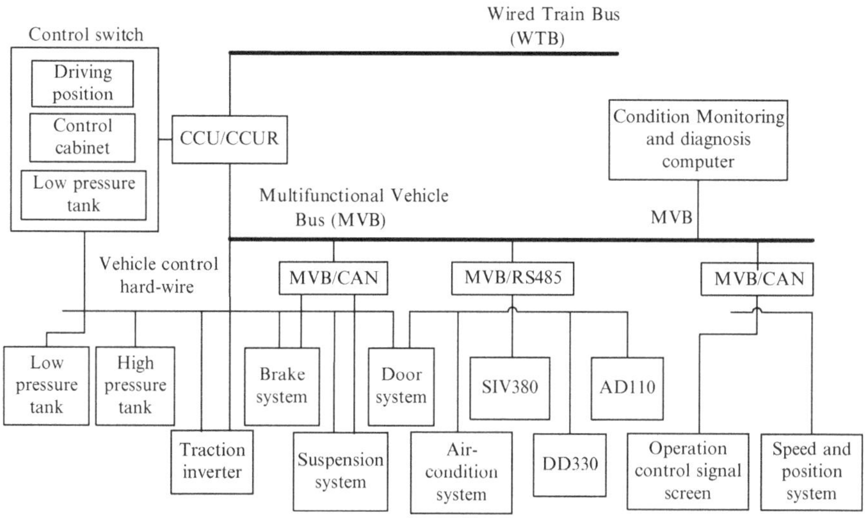


**Fig. 4.1** Structure diagram of Maglev train control and diagnosis system

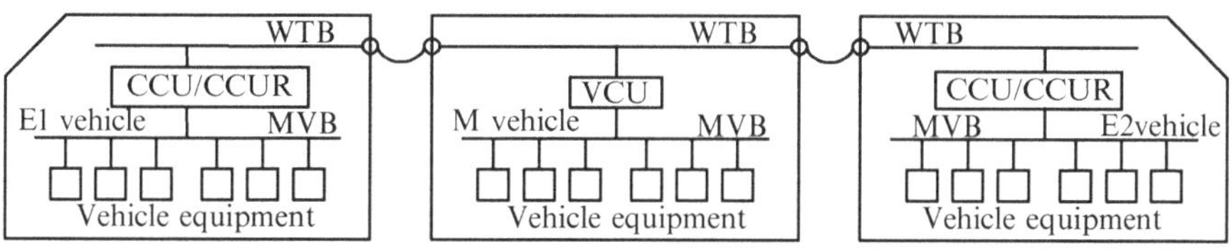


**Fig. 4.2** Composition of train control and diagnosis network

- *Information channel*
  Wired train bus, multifunctional vehicle bus, train line, and vehicle line are used to transmit the control instructions and control parameters and monitor the status of the train.

#### 4.2.1.2 Control and Diagnosis Network

The train control and diagnosis network is shown in Fig. 4.2, which is composed of the following parts:

- *Wired train bus*
  It is distributed through the whole train, connects both CCU/CCUR and VCU of every train, and is used to deliver information of the train.
- *CCU/CCUR and VCU*
  As the gateway between the wired train bus and multifunctional vehicle bus, it is used to complete the train control and vehicle control.

- *Multifunctional vehicle bus*
  It can connect the equipment inside the vehicle and is used to deliver the vehicle information.
- *Onboard equipment*
  They can connect with the multifunctional vehicle bus, exchange the information with CCU/CCUR and VCU via the network, and exchange the status information with the multifunctional vehicle bus via RS-485 or CAN bus.

#### 4.2.1.3 Hardwired Control

In the normal conditions, the onboard control system is always able to transmit control instructions of the bridge to CCU, or transmit them to every VCU through the wired train bus (WTB) after logic computation of CCU, and then to specific onboard equipment through multifunctional vehicle bus (MVB). In the case of faults of CCU or VCU, or complete paralysis of the train control network, the control mode will be ineffective, and the control instructions could not be transferred to the vehicle equipment. A heavy influence will be generated on safe operation. Thus, the hardwire control is required to add to the key equipment to ensure that the train control system can operate in the case of any fault.

### *4.2.2 Data Communication Network of the Control and Diagnosis System*

Maglev train control and diagnosis system adopts distributed computer network system as shown in Fig. 4.3.

All the control devices distributed in one carriage are networked by means of the vehicle-level bus. And the main control unit nodes distributed in different carriages are networked by means of the train-level bus and installed in the train control and diagnosis center, which is located at the front carriage and the rear carriage in Maglev train. They can be displayed on the display screen of driver's console. The operation status of controlled equipment in the train is optionally shown by means of display screen on driver's console of the end carriage, which can realize the comprehensive control and diagnosis of the whole train [9, 10].

Maglev train control and diagnosis system can be divided into three levels in terms of its structure, namely, train level, vehicle level, and subsystem level. The two-level networks, namely, the train-level network and the vehicle-level network, are connected together by the gateway to constitute a tree network system. In addition, Maglev train control and diagnosis system can be divided, in terms of its functions, into different functions, such as power supply control, suspension control,

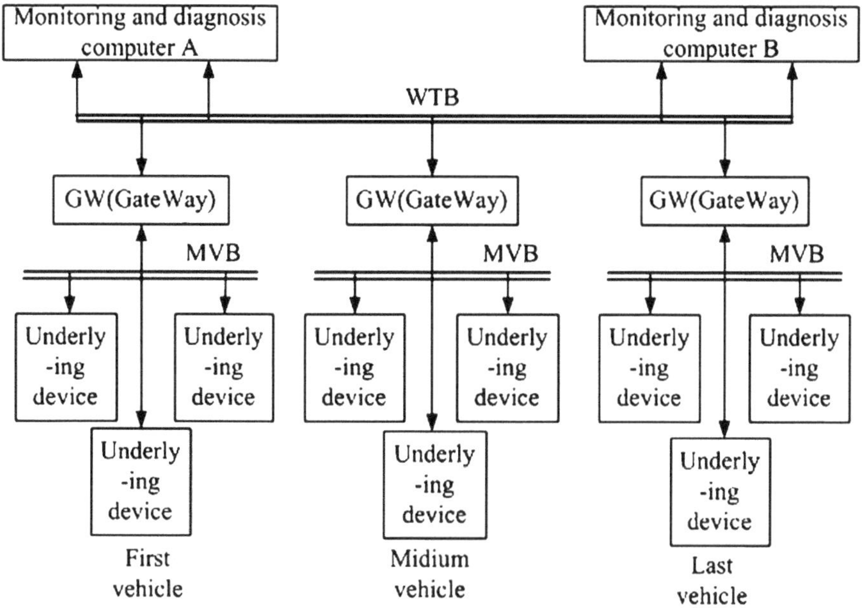


**Fig. 4.3** Block diagram of overall network composition of Maglev train control and diagnosis system

traction control, brake control, electric control of carriages, status monitoring of related equipment, and fault diagnosis [11].

The train communication network, which includes the train control and diagnosis communication network, can connect every level and every unit of the train control and diagnosis system to exchange and share the channel of information in the whole train and achieve information interaction in the environment of the whole train. In Maglev train, the various models, such as Profibus + CAN structure, industrial Internet + CAN structure, and WTB (wired train bus) + MVB (multifunctional vehicle bus) structure, have been adopted as train-level bus/vehicle-level bus in succession. In this section, WTB + MVB bus is adopted to meet IEC-61375 standard in the practical Maglev train.

### 4.2.3 *Hierarchy of Onboard Diagnosis System*

The overall structure of the onboard fault diagnosis system of Maglev train is shown in Fig. 4.3. The onboard fault diagnosis system is divided into the following three levels in terms of its structure.

#### 4.2.3.1 Equipment-Level Fault Diagnosis

The equipment control device can monitor its status and compare the status with the related standard values to achieve the self-fault diagnosis and input the diagnostic data into the control unit in accordance with the predetermined codes on this basis. The control unit can transmit the fault information to next diagnosis level.

#### 4.2.3.2 Vehicle-Level Fault Diagnosis

It involves the front carriage and the middle carriages. The nodes of every vehicle acquire, classify, and assess the fault diagnostic data via the multifunctional vehicle bus and then store these data information in a certain way and meanwhile transmit these fault information to the train-level diagnosis computer for the fault listing in terms of the predetermined single vehicle diagnosis parameter codes.

#### 4.2.3.3 Train-Level Fault Diagnosis

The train-level diagnosis computer, which is installed on both end carriages, can acquire, classify, assess, and store the diagnosis results of the train faults via the wired train bus. In addition, the results of major fault diagnosis are shown on the display screen in both end carriages for the driver. The detailed information on the faults can be read from each control ports for the maintainers.

Meanwhile, the diagnosis in the abovementioned levels is provided with the functions of self-fault diagnosis, fault information storage, necessary self-fault elimination, and important fault transmission to the upper level.

The diagnosis results are input into the onboard microcomputer system in the operation process for judging and classifying related instructions, which are given to the train control to ensure the safe operation of Maglev train. In addition, the results of fault diagnosis are sent to the train status data storage device to provide necessary information for maintenance in the process of onboard maintenance and overhaul of the train.

## 4.3 Comprehensive Assessment Algorithm of Onboard Faults Based on Fuzzy Comprehensive Assessment

Based on the comprehensive fault assessment methods of wheel-track railway, the fuzzy comprehensive assessment method of the system faults is adopted [9]. In addition, considering the large amount of onboard electric equipment of Maglev train and their close relations, the fuzzy comprehensive fault assessment is investigated with the object of onboard electric equipment.

### 4.3.1 *Modeling Principle for Fuzzy Comprehensive Assessment of Faults*

The fuzzy comprehensive assessment method refers to making the comprehensive assessment of membership grade of the assessed objects from multiple factors by applying the principle of synthesis of fuzzy relation [12, 13]. By utilizing the mathematic rule built on the concept of fuzzy set, the fuzzy membership function to express and process the difficult concepts can be adopted to keep precise. The fuzzy comprehensive fault assessment means to analyze and estimate the influence, in terms of the causal relationship of different extents between the reason for every fault and the fault units. By applying the principle of fuzzy transformation and corresponding assessment principle, the fault level of the system can be accessed on the basis of giving comprehensive consideration to all fault units. This method, as one kind of method that integrates the fuzzy theory and the membership principle, can adopt the concept of membership in fuzzy mathematics to represent the fuzzy relation between the fault symptoms and the symptom domain and introduce one kind of important coefficient, namely, weight, to solve the problem of comprehensive assessment of diversified faults. By the fuzzy operation, the method can not only reduce the requirement for information accuracy but also solve the problem of nonlinear mapping by means of multilevel assessment.

The fuzzy comprehensive fault assessment method of Maglev train can comprehensively consider all the factors that affect the assessment results and distribute a certain weight for these factors. In addition, comprehensive assessment results can be provided through the fuzzy matrix operation to decide the fault level of the train system and prompt the driver or the automatic driving system of the train to take measures against the disastrous accidents.

#### 4.3.1.1 Single-Level Fuzzy Comprehensive Fault Assessment Model

First, several related concepts in building fuzzy assessment models are introduced.

Assessment factor set $U = \{u_1, u_2, \dots u_n\}$ is a set of all factors that influence the assessment results $u_i$, which represents the underlying equipment or component-level faults. Assessment level set is $V = \{v_1, v_2, \dots v_m\}$. The assessment results are divided into $m$ levels in terms of the specific requirements. And the system faults can be divided into three levels, in particular, $V = \{$minor fault, general fault, major fault$\}$.

Factor assessment matrix $\underset{\sim}{R} \in F(U \times V)$ is below.

$$\underset{\sim}{R} = \begin{bmatrix} f(u_1) \\ f(u_2) \\ \vdots \\ f(u_n) \end{bmatrix} = \begin{matrix} \\ u_1 \\ u_2 \\ \vdots \\ u_n \end{matrix} \begin{matrix} v_1 & v_2 & & v_m \\ \left[\begin{matrix} r_{11} \\ r_{21} \\ \vdots \\ r_{n1} \end{matrix}\right. & \begin{matrix} r_{12} \\ r_{22} \\ \vdots \\ r_{n2} \end{matrix} & \begin{matrix} \cdots \\ \cdots \\ \cdots \\ \cdots \end{matrix} & \left.\begin{matrix} r_{1m} \\ r_{2m} \\ \vdots \\ r_{nm} \end{matrix}\right] \end{matrix} \tag{4.1}$$

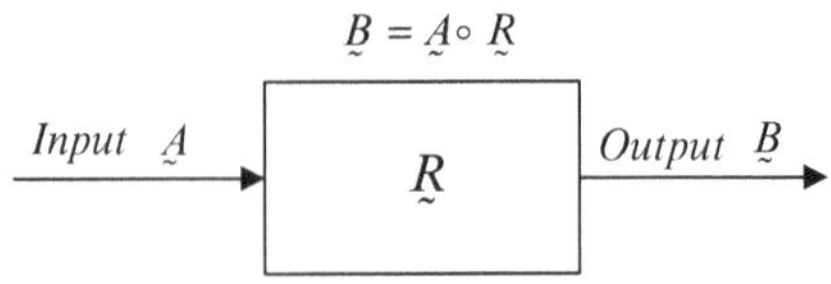


**Fig. 4.4** Block diagram of single-level fuzzy comprehensive assessment process

It represents the fuzzy relation matrix of from $U$ to $V$, in which $f(u_i)$ represents the assessment results obtained from the assessment of factor $u_i$. $r_{ij}$ represents the portion of the result of assessing factor set $u_i$ in assessment set $v_j$ level.

Weight set is $\underset{\sim}{A} = \{a_1, a_2, \ldots a_n\}$, in which $a_i$ $(0 \leq a_i \leq 1)$ is referred to the important coefficient of factor $u_i$. The elements in weight set meet the condition of $\sum_{i=1}^{n} a_i = 1$. When the fuzzy relation matrix $\underset{\sim}{R}$ and the weight set $\underset{\sim}{A}$ are determined, Formula (4.2) can be used for the fuzzy comprehensive assessment. The single-level fuzzy comprehensive assessment can be expressed by the block diagram as shown in Fig. 4.4.

$$\underset{\sim}{B} = \underset{\sim}{A} \circ \underset{\sim}{R} \in F(V) \text{ where } \underset{\sim}{B} = \{b_1, b_2, \ldots, b_m\} \tag{4.2}$$

The comprehensive assessment result $\underset{\sim}{B}$ is the fuzzy subset on the assessment set $V$, and $b_j$ is the membership degree of level $v_j$ for fuzzy subset $\underset{\sim}{B}$ obtained from the comprehensive assessment. As the fuzzy comprehensive assessment set, $\underset{\sim}{B}$ is also the fuzzy subset on assessment set $V$, and it is required to process the assessment index further in various prevailing methods including the maximum membership method, the weighted average method, the fuzzy distribution method, and so on. Based on the method features, they should be adopted properly in terms of different assessment objects.

Under the generalized fuzzy operation, the elements of $\underset{\sim}{B}$ can be expressed as follows:

$$b_j = \left(a_1 \dot{*} r_{1j}\right) \overset{+}{*} \left(a_2 \dot{*} r_{2j}\right) \overset{+}{*} \ldots \overset{+}{*} \left(a_n \dot{*} r_{nj}\right) \tag{4.3}$$

where $j = 1, 2, \ldots m$, $\dot{*}$ means the generalized fuzzy "and" operation, and $\overset{+}{*}$ means the generalized fuzzy "or" operation. The model can be abbreviated as $M$ $(\dot{*}, \overset{+}{*})$. Operated by the generalized "and" in the formula, the result is $a_i \dot{*} r_{ij}$ and $r*_{ij}$ is the membership of factor $u_i$ for assessment level $v_j$, if all comprehensive consideration to the influence of all factors is given. The generalized "or" operation $\overset{+}{*}$ means comprehensive disposal of the membership $r*_{ij}$ after every adjustment, where the reasonable result of comprehensive assessment can be obtained.

#### 4.3.1.2 Multilevel Fuzzy Comprehensive Fault Assessment Model

In the complicated system, there are many levels of factors that can affect the system faults, and they should be considered. In some complicated conditions, it is very difficult to use the single-level comprehensive assessment model to compare the priority order of the diagnosis objects in the system and let alone get any significant assessment result. Then, it will be difficult to distribute the fuzzy subset of important fault so that much information is drowned after operation. The assessment system will neglect the "minor" faults possibly to cause significant potential safety problems of the system. Thus, the fault set can be classified into several kinds in terms of some attributes. If the comprehensive assessment of every such kind is made, a higher level of the comprehensive assessment on basis of the various assessment results, which adopts the multilevel fuzzy comprehensive assessment model, will be made. Therefore, three-level assessment model is taken as an example to describe the assessment steps of the multilevel assessment model, which can be expressed by the block diagram as shown in Fig. 4.5.

Let $U = \{u_1, u_2, \ldots u_N\}$ represent the Layer I (Top) assessment factor set. The corresponding weight set is denoted by $\underset{\sim}{A} = \{a_1, a_2, \ldots, a_N\}$, and the fuzzy constraint relation is denoted by $\underset{\sim}{R}$. Let $U_i = \left\{u_{i1}, u_{i2}, \ldots, u_{in_j}\right\}$ represent Layer II (mid-layer) assessment factor set, $\underset{\sim i}{A}$ represent the corresponding weight, and $\underset{\sim i}{R}$

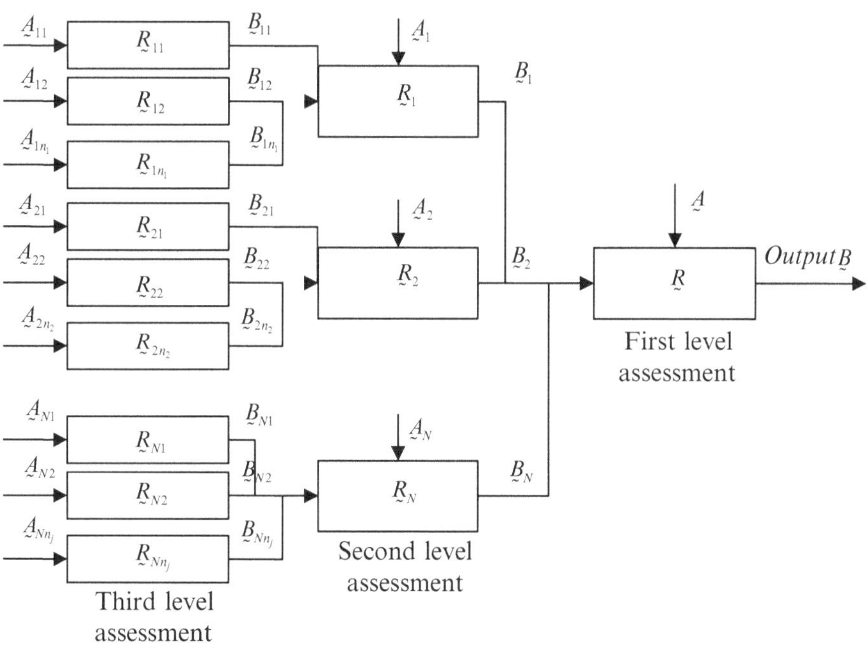


**Fig. 4.5** Block diagram of multilevel fuzzy comprehensive assessment process

represent the corresponding fuzzy constraint relation. Let $U_{ij} = \{u_{ij1}, u_{ij2}, \ldots u_{ijl}\}$ represent Layer III (bottom layer) assessment factor set, $\underset{\sim ij}{A}$ represent the corresponding weight set, and $\underset{\sim ij}{R}$ represent the corresponding fuzzy constraint relation. The specific assessment steps are shown as follows:

1. Level III comprehensive assessment

Compute the membership value of equipment faults $S\left(U_{ij}\right) = (S_1, S_2, \ldots S_l)$ based on the corresponding detection equipment status module of the system. Thus, the fault assessment result of Level III assessment factor set $U_{ij}$ can be obtained, namely, $\underset{\sim ij}{B} = \left[\underset{\sim ij}{A} \times S\left(U_{ij}\right)\right] \circ \underset{\sim ij}{R}$. The next level of the fuzzy relation matrix $\underset{\sim i}{R}$ is determined to be listed below:

$$\underset{\sim i}{R} = \left[\underset{\sim i1}{B} \; \underset{\sim i2}{B} \; \ldots \; \underset{\sim im}{B}\right]^{\mathrm{T}} \tag{4.4}$$

2. Layer II comprehensive assessment

Assess the factor set $U_i$ to obtain $\underset{\sim i}{B}$, and determine the next layer of fuzzy relation matrix $\underset{\sim}{R}$ to be listed below:

$$\underset{\sim i}{B} = \underset{\sim i}{A} \circ \underset{\sim i}{R} \tag{4.5}$$

$$\underset{\sim}{R} = \left[\underset{\sim 1}{B} \; \underset{\sim 2}{B} \; \ldots \; \underset{\sim N}{B}\right]^{\mathrm{T}} \tag{4.6}$$

3. Layer I comprehensive assessment

Assess the factor set $U$ to obtain the assessment result $\underset{\sim}{B}$, and determine the system fault level by adopting the maximum membership method as follows:

$$\underset{\sim}{B} = \underset{\sim}{A} \circ \underset{\sim}{R} \qquad j = \left\{ j \,\middle|\, \max\left(\underset{\sim j}{b}\right), \underset{\sim j}{b} \in \underset{\sim}{B} \right\} \tag{4.7}$$

#### 4.3.1.3 Multilevel Fuzzy Comprehensive Fault Assessment Method

This section is aimed at researching the fault level assessment of Maglev train system composed of numerous factor levels. It requires the comprehensive consideration of all kinds of fault factors of the equipment. However, the severity order of the equipment faults plays different influences on the train system, and even it is possible to integrate the minor faults of multiple equipment and elements, which may become a harmful fault for the train system. For this reason, $M$ (•,+) for the comprehensive assessment model is adopted; meanwhile the influence of major faults and minor faults on the system and the integrity of the train system are considered. In addition, referring to the majority practice of the present track transit, the maximum membership rule for assessment of fault level in allusion to the result of comprehensive assessment $\underset{\sim}{B}$ is adopted.

## 4.3.2 Building of Multilevel Fuzzy Comprehensive Assessment Model for Maglev Train System Faults

In this section, with the research object of 2-train marshaling practical Maglev train and Tangshan line in China, the multilevel fuzzy comprehensive assessment model for Maglev train system faults based on the modeling principle of the comprehensive assessment model is built [14–16]. Considering the large amount and multilevels of the onboard electric equipment and their close relation, the following fuzzy comprehensive fault assessment cases are analyzed by virtue of the examples of onboard status monitoring and comprehensive fault assessment. The onboard status monitoring and assessment objects can cover main parts of the electric system of the vehicle and include the random and irregular track that can affect the stability of suspension. The main structure of the equipment is shown in Fig. 4.6.

### 4.3.2.1 Division of Assessment Set

The faults of Maglev train are divided into three levels according to the practice of railway, metro, and light rail.

Level I fault of train – the train can continue operation since the fault will not affect the operation of the train. It requires the overhauling in the maintenance base after operation in the current day. Such faults are classified as the minor faults.

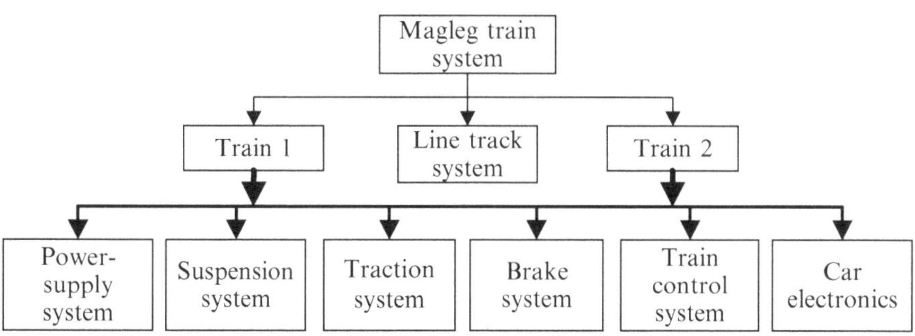


**Fig. 4.6** Main composition of train system

Level II fault of train – a functional fault occurs. The train must reduce its speed by reducing or clearing 1/4 of the power as prompted by this system. Then, the train should be overhauled after arriving at the terminal station. Such faults are classified as the general faults.

Level III fault of train – a major fault occurs. The train must reduce its speed by reducing or clearing 1/2 of the power as prompted by this system. After the Maglev train arrives at the next station, all passengers get off, and the train is overhauled in the maintenance base. Such faults are classified as the major faults.

The fault level of the equipment or parts of Maglev train is specified as follows:

Level A fault of equipment – some faults exist or have occurred despite of normal main functions.

Level B fault of equipment – the fault can affect the realization of major functions.

Level C fault of equipment – the major fault can damage the realization of major functions.

#### 4.3.2.2 Level Division of Factor Set

In practical commercial operation, Maglev train mostly runs in the way of multi-train marshaling. Owing to the design redundancy of power supply, the traction, and brake subsystem, in the process of multi-train coupling operation, these faults of subsystem parts or equipment of single train in the case of different marshaling play different influences on the train-level faults. Every carriage in Maglev train has consistent power supply systems. The fault of invalidated DD330V power supply will affect the suspension [12]. The fault of invalidated AD110V will affect the control of the whole train. When the marshaling operation of train is designed, the DD330V and AD110V of every carriage are adopted to connect in parallel, respectively, in the whole train. The method of expansion power supply for AC380V power supply of every carriage is used, that is to say, it is taken into account that the power supply DD330V, AD110V, and AC380V of other carriages can supply power for the single power supply fault carriage. Therefore, the faults of DD330V, AD110V, and AC380V power supply show different influences on trains with multi-train marshaling and single train, the traction system, and the brake system. If the requirements for flexibility of the abovementioned related redundancy design and carriage marshaling are considered, the model structure and parameters for the comprehensive train-level fault assessment will be adjusted along with different marshaling quantities. In terms of the functional modules, the comprehensive assessment model of Maglev train system fault is built according to the conditions of actual two-train marshaling for Maglev train.

### *4.3.3 Application and Analysis of Fuzzy Comprehensive Assessment of Maglev Train System Faults*

#### 4.3.3.1 Framework of Fuzzy Comprehensive Fault Assessment Process

Figure 4.7 is the framework of comprehensive fault assessment process of the train system. According to the actual two-train marshaling condition at present, the train-level comprehensive fault assessment system adopts the four-level fuzzy comprehensive assessment method for level-by-level assessment and finally determinates the fault level of train based on the maximum membership rule [17, 18].

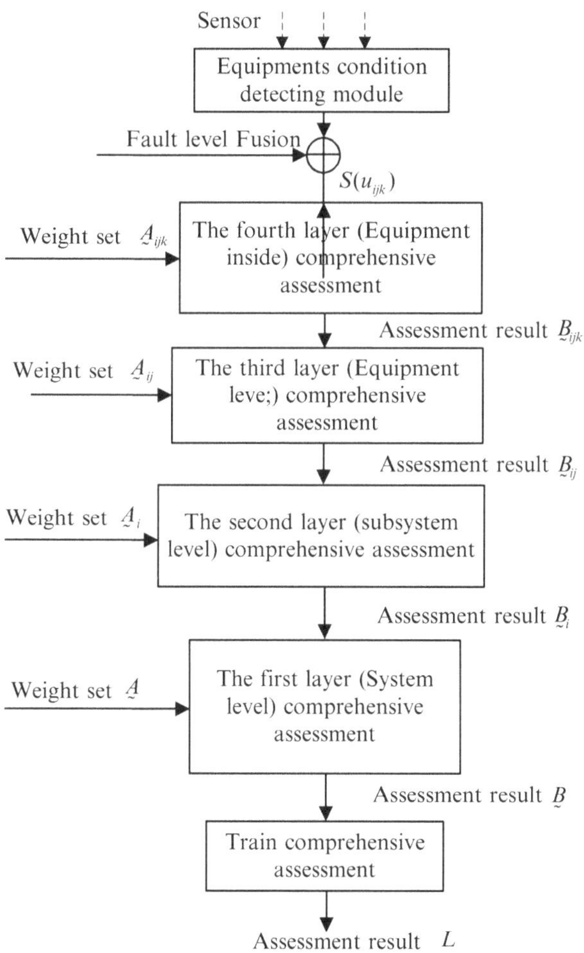


**Fig. 4.7** Fuzzy comprehensive assessment process of train faults

The specific algorithm process is described as follows:

1. Determine the level fusion or division of the fault factor of equipment

Every fault is divided into five levels. If Level 1 factor fault of equipment or parts is $u_{ijk}$, $S\left(u_{ijk}\right) = \left(1\ 0\ 0\ 0\ 0\right)$, suppose $S\left(u_{ijk}\right) = (s_1, s_2, s_3, s_4, s_5)$. Level $l$ $(1 \le l \le 5)$ factor fault of equipment or parts $u_{ijk}$ occurs, thus,

$$s_i = \begin{cases} 0; (i \neq l) \\ 1; (i = l) \end{cases}. \tag{4.8}$$

2. Level IV comprehensive assessment

It refers to the comprehensive assessment of the severity of equipment or parts' faults. Based on the above discussion, the factor level constraint vector $S(u_{ijk})$ of equipment or parts $u_{ijkl}$ can be obtained. Thus, the assessment result of equipment or parts' faults can be obtained, $u_{ijk}$ $\underset{\sim ijk}{B} = \left[\underset{\sim ijk}{A} \times S\left(u_{ijk}\right)\right] \circ \underset{\sim ijk}{R}$, where "×" operation represents the point multiplication and "∘" represents the synthetic operation. It can be made by ordinary multiplication and addition as defined by $M$ (•, +) model.

3. Level III comprehensive assessment

The step is finished in the subsystem. The assessment formula is shown as follows.

$$\underset{\sim ij}{B} = \underset{\sim ij}{A} \circ \underset{\sim ij}{R} \tag{4.9}$$

$$\underset{\sim ij}{R} = \left[\underset{\sim ij1}{B}, \underset{\sim ij2}{B}, \ldots \underset{\sim ijn}{B}\right]^{\mathrm{T}} \tag{4.10}$$

where "T" represents the matrix transposition and $n$ represents the factor subset. $u_{ij}$ has totally $n$ subfactors, $i = 1, 2, \ldots 6$.

4. Level II comprehensive assessment

The assessment is aimed at the functional modules. The assessment formula is shown as follows:

$$\underset{\sim i}{B} = \underset{\sim i}{A} \circ \underset{\sim i}{R} \tag{4.11}$$

$$\underset{\sim i}{R} = \left[\underset{\sim i1}{B}, \underset{\sim i2}{B}, \ldots \underset{\sim in}{B}\right]^{\mathrm{T}} \tag{4.12}$$

where "$T$" represents the matrix transposition and $n$ means the factor subset $U_i$ has totally $n$ subfactors, $i = 1, 2, \ldots 6$.

5. Level I comprehensive assessment

The assessment is finished for the train system fault level. The assessment formula is shown as follows:

$$\underset{\sim}{B} = \underset{\sim}{A} \circ \underset{\sim}{R} \tag{4.13}$$

$$\underset{\sim}{R} = \left[ \underset{\sim 1}{B} \; \underset{\sim 2}{B} \; \cdots \; \underset{\sim 6}{B} \right]^{\mathrm{T}} \tag{4.14}$$

6. Determine the assessment result

Finally, the vector $\underset{\sim}{B} = (b_1, b_2, b_3)$ is computed. The fault level of train system is determined based on the maximum membership rule as follows:

$$j = \left\{ i \middle| \max(b_i), b_i \in \underset{\sim}{B} \right\} \tag{4.15}$$

#### 4.3.3.2 Experimental Analysis

If the following faults occur in the whole train system, the onboard monitoring and diagnosis system can detect that the output overvoltage and the input overvoltage (2 Level B faults) occur in the DD330V power supply system of Front Carriage 2. It can be obtained that one Level 3 factor fault occurs to the DD330V power supply of the Front Carriage 2. Thus, the factor level constraint vector $S(u_{151})$ inside the DD330V equipment $u_{151}$ can be obtained as follows:

$$S(u_{151}) = (0\ 0\ 1\ 0\ 0) \tag{4.16}$$

The specific level assessment process of comprehensive fault assessment is the same as Condition 1 above, meanwhile the two-train marshaling condition is considered: hence, $\underset{\sim}{B} = \underset{\sim}{A} \circ \underset{\sim}{R}$ can be obtained finally as follows:

$$\begin{aligned} \underset{\sim}{B} &= \underset{\sim}{A} \circ \underset{\sim}{R} \\ &= [0.15{*}0.35{*}0.17{*}0.13 \quad 0.15{*}0.55{*}0.17{*}0.13 \quad 0.15{*}0.1{*}0.17{*}0.13] \\ &= [0.00116025 \quad 0.00182325 \quad 0.0003315] \end{aligned} \tag{4.17}$$

By applying the maximum membership rule to the comprehensive assessment result $\underset{\sim}{B}$, it can be found that Level II general fault occurs for the train system.

If the following faults occur in the whole train system, the onboard monitoring and diagnosis system can not only detect that the output over current occurs in the

AD110V power supply of Front Carriage 1 (one Level C fault) but also detect that the output overvoltage and the input voltage occur in the DD330V power supply system of Front Carriage 2 (2 Level B faults). Likewise, Level 4 factor fault occurs in the AD110V power supply of Front Carriage 1. One Level 3 factor fault occurs in the DD330V power supply of Front Carriage 2. Therefore, the factor level constraint vector $S(u_{111})$ inside AD110V equipment $u_{111}$ and the factor level constraint vector $S(u_{151})$ inside DD330V equipment $u_{151}$ can be listed as follows:

$$\begin{aligned} S\left(u_{111}\right) &= \left(0\;0\;0\;1\;0\right) \\ S\left(u_{151}\right) &= \left(0\;0\;1\;0\;0\right) \end{aligned} \tag{4.18}$$

The specific level assessment process of the comprehensive fault assessment is the same as the condition above; meanwhile the two-train marshaling condition is considered; hence, $\underset{\sim}{B} = \underset{\sim}{A} \circ \underset{\sim}{R}$ can be obtained finally as follows:

$$\begin{aligned} \underset{\sim}{B} &= \underset{\sim}{A} \circ \underset{\sim}{R} \\ &= 0.13 \times \begin{bmatrix} 0.17 * 0.15 * 0.35 \\ 0.0875 * 0.17 + 0.15 * 0.55 * 0.17 \\ 0.17 * 0.1625 + 0.15 * 0.1 * 0.17 \end{bmatrix}^{\mathrm{T}} \\ &= \begin{bmatrix} 0.00116025 & 0.00193375 + 0.00182325 & 0.00359125 + 0.0003315 \end{bmatrix} \\ &= \begin{bmatrix} 0.00116025 & 0.003757 & 0.00392275 \end{bmatrix} \end{aligned} \tag{4.19}$$

It can be seen that major fault occurs for the train system by applying the maximum membership rule to the comprehensive assessment result.

## 4.4 Comprehensive Assessment Method Based on EDA

### *4.4.1 Questions*

Considering the complicated structure of Maglev train system and much onboard electric equipment, it is very difficult to ensure that all equipment can keep working in normal condition at all times in the operation process [19–21]. In order to enhance the fault-tolerant performance of the system, the comprehensive influence of the part faults or the subsystem faults should be assessed correctly, which plays a significant role in instructing the train driver or the automatic control system to take corresponding measures to guarantee passengers' life safety.

The fuzzy comprehensive assessment is a method that is widely used to assess the wheel-track train fault level. Its advantages include simple structure, understandability, and high execution efficiency (computational efficiency), while its disadvantage is the difficult establishment and modification of fuzzy model parameters (important parameters and membership parameters). For the judgment of the fault level of Maglev train system, in particular, considering the numerous model parameters, it is not scientific or easy to develop and widely promote the method relying on experts merely. Moreover, in order to judge the reasonability of model parameters that have an important influence on the correctness of the final fault level judgment, it is required to research the automatic determination and the optimization method for the fuzzy assessment model parameters.

In recent years, the flourishing machine learning becomes a research focus. The cross of machine learning and other disciplines tends to play a great effect. Especially, the rapidly developing evolutionary learning development has provided with some important methods and tools for solving the problems of optimizing various complicated parameters. Based on the data driving and approximation modeling, the evolutionary learning can make uses of the actual observation data for the approximation modeling. By combining with the given model, the evolutionary algorithm for automatic optimization of model parameters in the modeling process can be used [22–24].

The estimation of distribution algorithm (EDA), as a kind of new random optimization algorithm of evolutionary computation, is a research hotspot in the present international computational intelligence [25–27]. EDA inherits the evolution idea of genetic algorithm. Being different from the major evolutionary computation methods that use cross operation and mutation operation as the necessary tools to generate new population, EDA can help to generate new population by estimating and simulating joint probability distribution of the selected individuals. This new evolution pattern can build models by virtue of the relation between the probability model and variable to solve the optimization related to multivariable effectively. The experimental analysis indicates that EDA shows a better performance than the general genetic algorithm in the process of solving high-dimensional problem. In short, it is very potential to apply EDA to solve complicated optimization problem in engineering and science.

It is difficult to determine the model parameters for the traditional fuzzy comprehensive judgment method on judging the fault level. In this section, a kind of fuzzy comprehensive fault judgment method based on EDA is proposed, which can improve the assessment accuracy via the evolutionary computation and retain the advantage of traditional fuzzy comprehensive judgment model, namely, easy to comprehend in form with high operation efficiency. Further, the fuzzy comprehensive fault judgment method based on EDA can have the complementary advantages and enhance correct assessment of comprehensive fault level of Maglev train.

### *4.4.2 Basic Theory of the Estimation of Distribution Algorithm*

The evolutionary algorithm is a kind of random heuristic algorithm advantaged by its ability to obtain the approximate optimal solution in the acceptable duration. As the evolutionary algorithm can provide one kind of encoding scheme for the whole parameter space, instead of dealing with the specific parameters of problems directly, and can realize the searching from a set of initial individuals, instead of some single initial individual, it has better robustness and global superiority.

However, there are two major problems as a whole for the traditional evolutionary algorithm shown as follows [28–30]:

1. The evolutionary performance of the algorithm is related with the parameters of algorithm itself. If the researchers have no experience in solving the problem of specific optimization, the selection of algorithm parameters is an optimal problem by itself.
2. It is extremely difficult for the algorithm to predict the population evolution if the correlation among different variables is not considered effectively in the selection and generation process of individuals.

EDA is one kind of new evolution patterns that does not involve the traditional genetic operation, such as cross and mutation. This kind of optimization technique can be used to build models by virtue of the relation between the probability of model and variables and control the evolutionary direction. With the joint probability distribution (density) of individuals, it can be iteratively selected every time. The basic algorithm is shown as follows:

(1) Initialize the population and generate random $N$ sets of individuals as the initial population.
(2) Repeat steps (3) and (4) till the termination condition is satisfied.
(3) Pick out better individuals as the dominant population by means of the population assessment and construct the population probability distribution.
(4) Sample and generate new population based on the probability distribution of the last generation of dominant population.

The probability model can be divided into the probability distribution of discrete domain and the probability density of continuous domain based on different definition domains of evolutionary variables.

At present, the EDA research of discrete domain is relatively mature, that is to say, the probability model can be expressed by Bayesian network. EDA can be divided into the following types in terms of the complication of probability mode:

1. Distribution estimation of the independent variables, for example, UMDA (univariate marginal distribution algorithm), PBIL (population-based incremental learning), CGA (compact genetic algorithm), etc.
2. Related distribution estimation of double variables, including MIMIC (mutual information maximization for input clustering), COMIT (combining optimizers with mutual information trees), etc.
3. Related distribution estimation of multivariable, including ECGA (extended compact genetic algorithm), FDA (factorized distribution algorithms), BOA (Bayesian optimization algorithm), etc.

The design of EDA algorithm in continuous domain is a relatively difficult process. The main reason is that every continuous variable has infinite values, which will possibly result in a very big searching space of optimization algorithm, and such contradiction becomes more prominent especially in the case of bigger variable dimension. As the result of more common continuous conditions in the actual application, many researches are made in this field, for example, the UMDA developed on the basis of UMDA, the PBILc developed on the basis of PBIL, MIMICcG related to double variables, EGNAee and EGNABGe, etc., related to multivariable.

### 4.4.3 Comprehensive Assessment of Maglev Train Faults Based on EDA

The optimizing process of fuzzy model parameters based on EDA is shown in Fig. 4.8.

The optimizing process is listed as follows.

1. Modeling

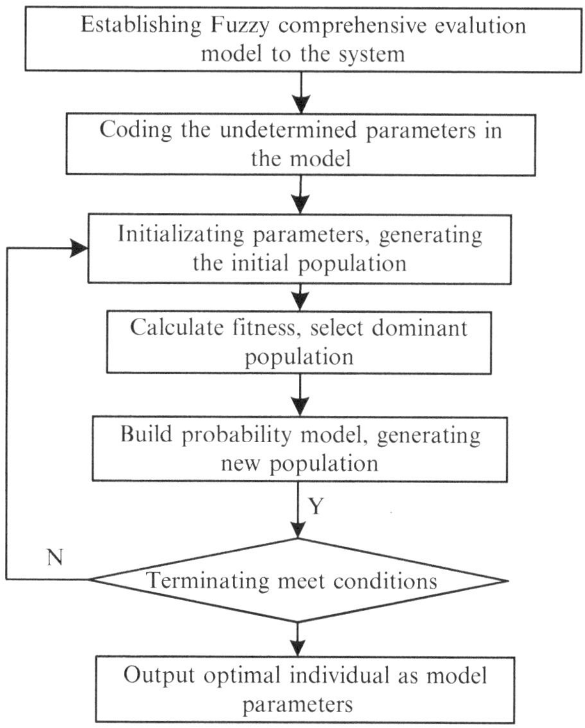


**Fig. 4.8** Flow chart of optimizing fuzzy model parameters by EDA

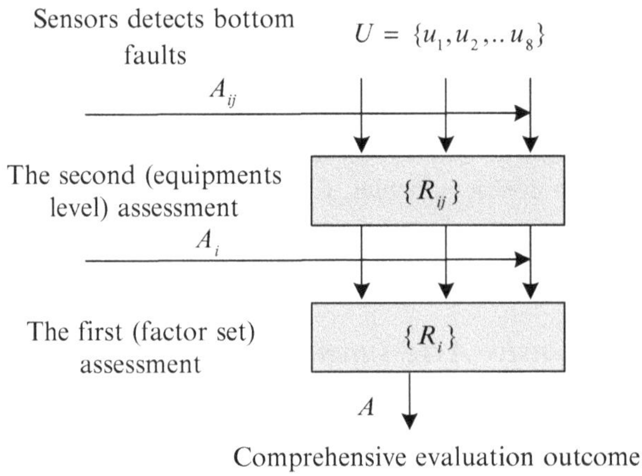


**Fig. 4.9** Schematic diagram of Level II fuzzy comprehensive assessment process of Maglev train suspension system

The suspension system of Maglev train is taken as the research object. The Level II comprehensive judgment model is built as shown in Fig. 4.9. After the simplification of suspension system, the influence of eight kinds of underlying faults will be specially studied.

$u_1$: Main circuit breaker of suspension
$u_2$: 280 VDC-DC
$u_3$: 110 VDC-DC
$u_4$: Diagnosis system
$u_5$: Bogie A
$u_6$: Bogie B
$u_7$: Bogie C
$u_8$: Bogie D

Each fault is divided into three levels. The comprehensive assessment level of top faults is also divided into three levels (minor fault, intermediate fault, major fault). The judgment factor set $U$ is $\{u_1, u_2, \ldots u_8\}$.

The dividing factor set $U$ is divided into four subsets, namely, $U_1 = \{u_1\}$, $U_2 = \{u_{2,} u_3\}$, $U_3 = \{u_4\}$, and $U_4 = \{u_5, u_6, u_7, u_8\}$, in terms of the relations among faults.

The Level II assessment weight set is $\left\{A_{ij} \middle| i = 1, 2, 3, 4\right\}$ and the membership matrix is $\left\{R_{ij} \middle| i = 1, 2, 3, 4\right\}$. The weight set of factor subsets of Level I assessment is $\left\{A_i \middle| i = 1, 2, 3, 4\right\}$ and the membership matrix is $\left\{R_i \middle| i = 1, 2, 3, 4\right\}$.

2. Encoding

The main task of encoding is to establish one-to-one correspondence between the solution space and the population individual space. It requires completeness, integrity, and nonredundancy.

The encoding can be divided into the binary encoding and the floating point encoding in terms of different research objects. For the research object in this section, the encoding is to encode the weight and membership parameters to be learned in the row vector. Owing to the range of parameters in [0 1], the floating point encoding should be adopted. Meantime, it is required to make full use of the known information in the encoding process, reduce unnecessary variables as much as possible, and choose the encoding scheme as simple as possible to increase the efficiency of evolutionary computation. There are totally 48 $(3+3\times3\times4+2+4+3)$ undetermined parameters for comprehensive judgment model. The weight of same level added to the membership equals to 1. So the number of parameters can be simplified to be $33((3–1)+(3–1)\times3\times4+(2–1)+(4–1)+3)$. Then, the solution space can be encoded in array of $1\times33$ dimension with the range of [0 1] for ensuring the nonredundancy.

3. Calculation of fitness

Fitness is a standard for judging the coincidence between the learning individual and actual results. The higher fitness indicates that the individual is more suitable for the actual environment better. As the definition of fitness function has not been required rigidly, it is allowed to use the reciprocal of general minimum mean square error for fitness standard or define the fitness function according to the minimum risk standard. For Maglev train, in order to make an object safety a top priority, there is no doubt that the minimum risk standard is more suitable.

$$\text{Fitness} = 1/\left(\lambda_{\|F(x,\alpha)-f(x)\|}\cdot\|F(x,\alpha)-f(x)\|^2\right) \tag{4.20}$$

Suppose the encoding form be $X=(X_1,X_2,\ldots,X_i,\ldots,X_n)$, the assessment individual generated randomly be $x=(x_1,x_2,\ldots,x_i,\ldots,x_n)$, and the comprehensive judgment model be $F(x,\alpha)$, where $x$ represents the fuzzy parameter or the coefficient to be optimized, $\alpha$ represents the input of underlying faults, $f(\alpha)$ represents the output of fault level, and $\lambda_{\|F(x,\alpha)-f(x)\|}$ represents the risk coefficient. The fitness function of minimum risk standard can be represented as follows:

$$\text{Fitness} = 1/\left(\lambda_{\|F(x,\alpha)-f(x)\|}\cdot\|F(x,\alpha)-f(x)\|^2\right) \tag{4.21}$$

$$\text{Fitness} = 1/\left(\lambda_{\|F(x,\alpha)-f(x)\|}\cdot\|F(x,\alpha)-f(x)\|^2+\varepsilon\right) \tag{4.22}$$

In order to avoid the divisor to be equal to zero, define $\varepsilon>0$, and thus the fitness function is rewritten as follows:

$$\text{Fitness} = 1/\left(\lambda_{\|F(x,\alpha)-f(x)\|}\cdot\|F(x,\alpha)-f(x)\|^2+\varepsilon\right) \tag{4.23}$$

4. Building the probability estimation model

The building of probability estimation model is the core of EDA and the key to realize the optimization of model parameters by EDA. The probability model is composed of two major parts, namely, structure and a set of local generalized probability density. Suppose $X = (X_1, X_2, \ldots, X_i, \ldots, X_n)$ be a set of random variables, where $X_i$ value is represented as $x_i$, $S$ represents the structure of $X$, $\rho(x)$ represents the local generalized probability density of $X$, and the variable set of father nodes of $X_i$ is $P\alpha_i^S$ $(i = 2, \ldots, n)$. As $X_i$ is only related to its father nodes, $X_i$ and $\{X_1, \ldots, X_n\} \Big| P\alpha_i^S$ become independent. So the probability distribution (density) can be decomposed as follows:

$$\begin{aligned}
&\rho(x) \\
&= \rho(x_1, \ldots, x_n) \\
&= \rho(x_1) \cdot \rho\left(x_2 \middle| x_1\right) \cdot \cdots \cdot \rho\left(x_i \middle| x_1, \ldots, x_{i-1}\right) \cdot \\
&\quad \cdots \cdot \rho\left(x_n \middle| x_1, \ldots, x_{n-1}\right) \\
&= \rho(x_1) \cdot \rho\left(x_2 \middle| P\alpha_2^S\right) \cdot \cdots \cdot \rho\left(x_i \middle| P\alpha_i^S\right) \cdot \\
&\quad \cdots \cdot \rho\left(x_n \middle| P\alpha_n^S\right) \\
&= \prod_{i=1}^{n} \rho\left(x_i \middle| P\alpha_i^S\right)
\end{aligned} \tag{4.24}$$

where $\theta_S = (\theta_1, \ldots, \theta_n)$.

Suppose that the local generalized probability density is related to the finite parameter set $\theta_S$ $(\theta_S \in \Theta_S)$, so the former equation can be represented as follows:

$$\rho\left(x \middle| \theta_S\right) = \prod_{i=1}^{n} \rho\left(x_i \middle| p\alpha_i^S, \theta_i\right) \tag{4.25}$$

where $\theta_S = (\theta_1, \ldots, \theta_n)$.

According to the definition of fuzzy comprehensive judgment model, the sum of the membership (weight parameters) of same level is equal to 1 and the model parameters are not independent. For EDA, however, the more complicated the correlation among variables is, the more complicated the probability estimation model is and the more difficult the calculation is. Meanwhile, the evolution will be affected because of the immature EDA researches that are related to multivariable at present.

By representing the memberships of the same level, such $\{x1, x2, x3\}$ as $\{x1, (1 - x1) \cdot x2, 1 - x1 - (1 - x1) \cdot x2\}$, the relationships among parameters are decomposed to make the variable independent so that the independent variable distribution method UMDAc can be used for the parameter optimization. And the probability distribution is simplified as follows:

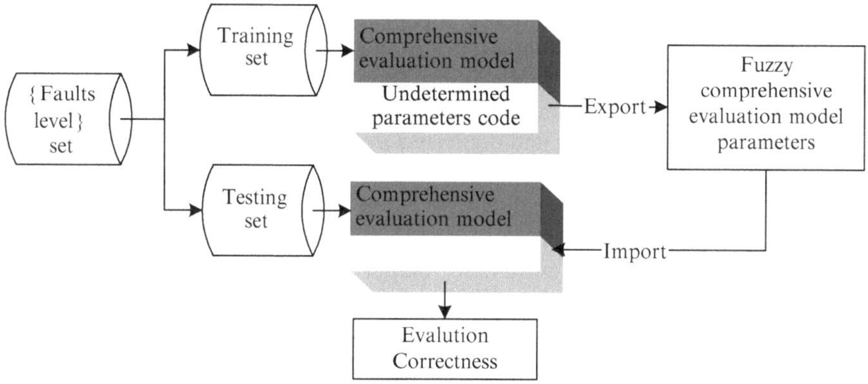


**Fig. 4.10** Experimental model of fuzzy comprehensive assessment based on EDA

$$\rho\left(x\middle|\theta_S\right) = \prod_{i=1}^{n} \rho\left(x_i, \theta_i\right) \tag{4.25}$$

In this way, the evolutionary is greatly simplified.

5. Training and testing model

The data training and testing process shown in Fig. 4.10 is mainly adopted in the experimental design.

The data of train operation faults in the form of fault level is collected and organized, some parts of which are used for training and the other for testing. After encoding the undetermined parameters in the sample, the training set is used for training and after the training, the optimized parameters are input into the comprehensive assessment model and the testing set for testing is used to assess the effect of evolutionary method.

In addition, although the random way is adopted for initializing population under general evolutionary algorithm, the individuals that are set roughly and artificially can be added to accelerate convergence of the algorithm. When it is required to amend the parameters because of former erroneous judgment, it is allowed to add the parameter individuals that are obtained from previous evolution to the initial population.

### 4.4.4 Performance Test and Comparison

This section will test the influence of the comprehensive fault judgment system parameters on the performance based on EDA and compare the assessment of model performance obtained based on the EDA, GA, and other machine learning algorithms.

For testing the influence of parameters, the correlation among variables is not considered in encoding. For comparing with other methods, the correlation will be decomposed as the above with the UMDAc probability model.

According to the operation record of Maglev train, 400 sets of samples that are used for training are collected and organized with the sample distribution as shown in Fig. 4.11. Hundred sets of samples are used for testing and the sample distribution is shown in Fig. 4.12.

In this figure, there are totally nine sub-histograms to describe the corresponding sample distribution of input attribute or output attribute respectively. The above-mentioned input includes eight attributes, namely, main circuit breaker of the suspension, 280 VDC-DC, 110 VDC-DC, diagnosis system, Bogie A, Bogie B, Bogie C, and Bogie D, and the output is the fault assessment level. In the distribution diagram, the black, the gray, and the light gray represent the quantity of Level 1 fault, Level 2 fault, and Level 3 fault of the train suspension system, respectively. For the sub-diagram of input attributes, these three colors, respectively, represent the quantity of corresponding fault level in the input attributes.

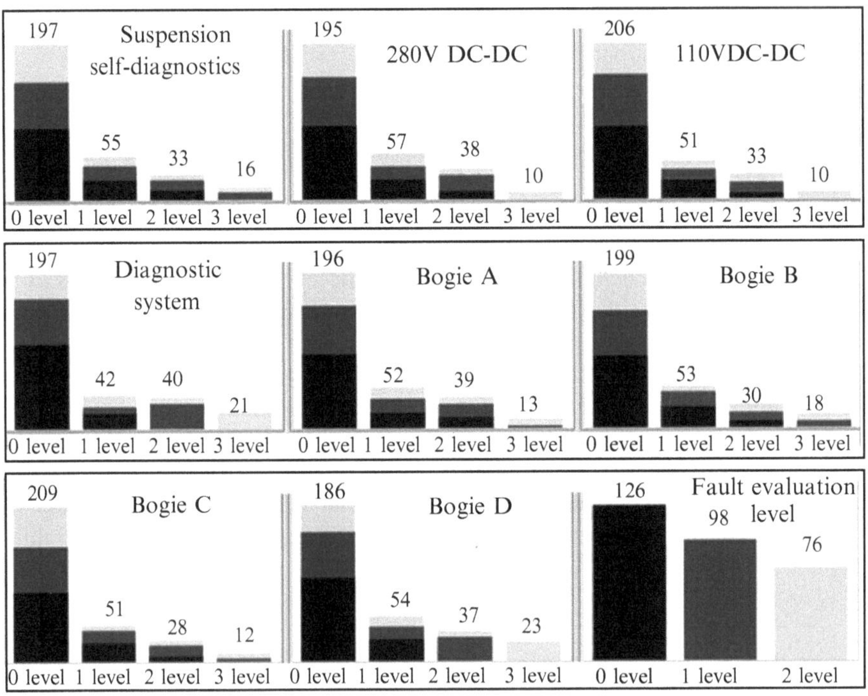


**Fig. 4.11** Distribution of sample data set used for 300 sets training

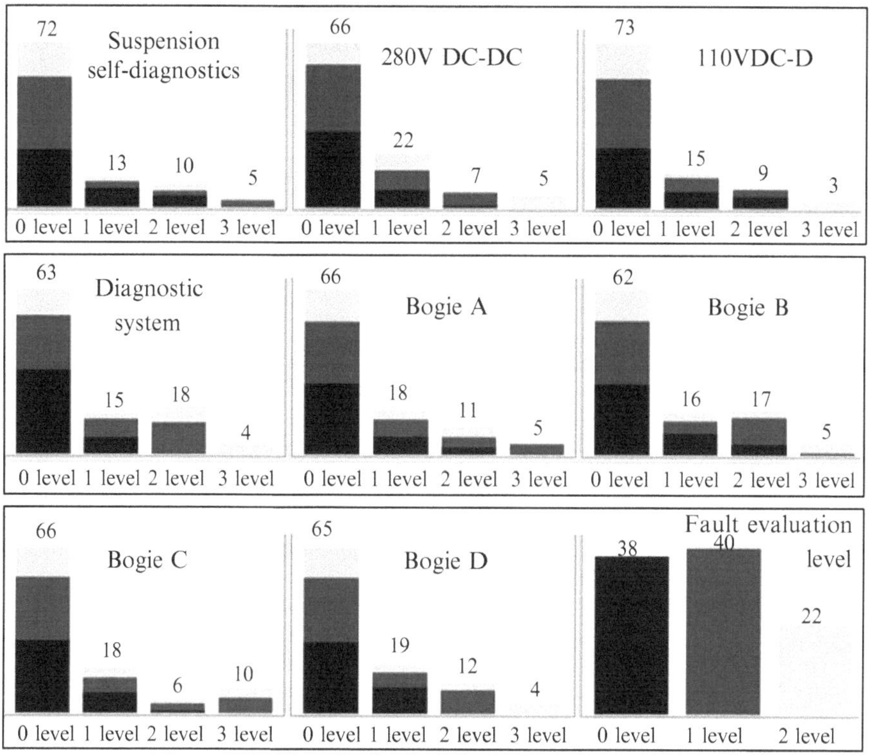


**Fig. 4.12** Sample data set used for testing, totally 100 sets

#### 4.4.4.1 Influence Test of Comprehensive Judgment System Parameters Based on EDA

The parameters of EDA include the population quantity, the truncating probability, the maximum iterating times, etc., in which the population quantity and the truncating probability are the key parameters of EDA.

1. Influence of the population quantity on EDA

Set the truncating probability to 0.2 and the maximum iterating times to 200. Change the quantity of initialized population, and then the training accuracy–learning times curve can be obtained as shown in Fig. 4.13, in which the population represents the quantity of population.

From Fig. 4.13, it can be seen that the judgment accuracy increases rapidly with the training times, though the random selection of initialized population results in the slightly different initial accuracy of each training process. When the training times reach 80, the training results keep stable at a high accuracy. The training accuracy shows a rising trend with the increasing quantity of population on the

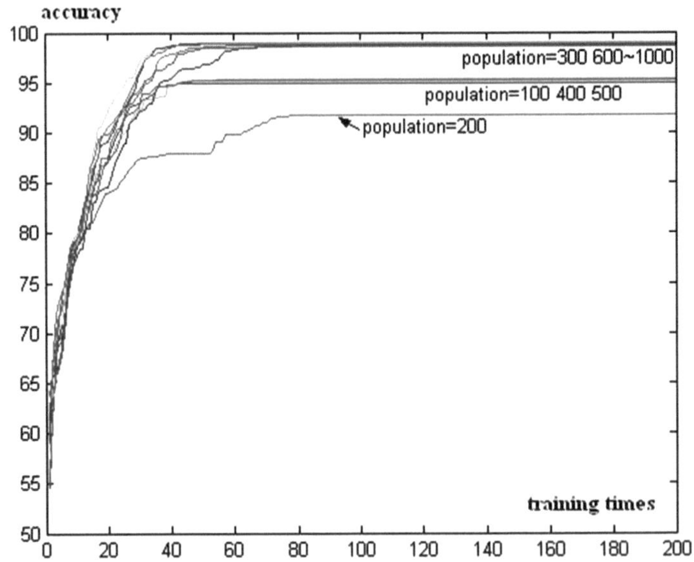


**Fig. 4.13** Training times–accuracy relationship curve for changing population quantity

whole. When the quantity of population is more than a certain value (600), the final result of the training will reduce greatly with the influence of random factor and keep stable within a range of high accuracy (99 %).

2. Influence of truncating probability on EDA

Set the population quantity to 300 and the maximum iterating times to 80. Change the truncating probability of each training process and then the training accuracy–learning times curve can be obtained as shown in Fig. 4.14. The truncation is used to represent the truncating probability.

Figure 4.14 shows the influence of truncating probability on the judgment results. It can be seen from the truncating probability and final accuracy curve, Truncation $\in [0.2, 0.3]$, the higher the accuracy is, the further it deviates from this range and the more obvious the trend of decreasing accuracy is. Such phenomenon can be understood in theory, for within the range of Truncation $\in [0.2, 0.3]$. The dominant population can represent the trend of the best optimal individual. Too low truncating probability indicates a big random element, and whereas too high truncating probability is difficult to show the probability distribution of the dominant population.

The experimental results show that EDA can converge rapidly to a smaller error range. With the increase of population quantity, the effect shows a tendency to stable optimum value, and the comprehensive fault judgment system based on EDA can obtain very high accuracy.

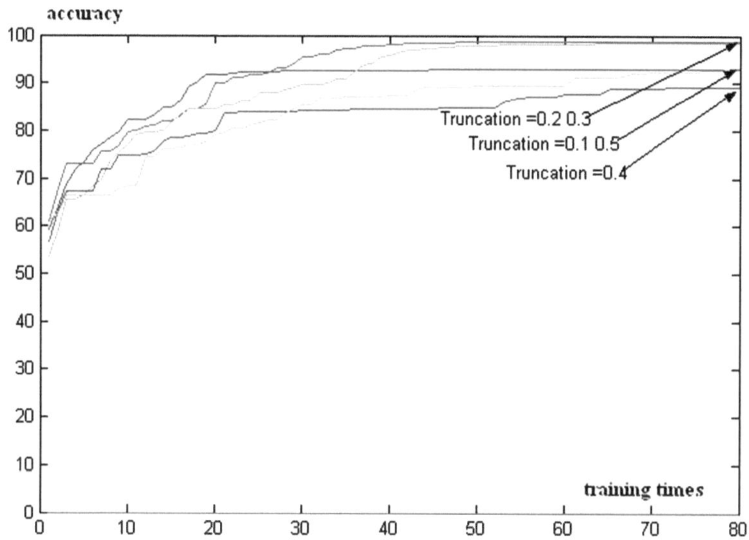


**Fig. 4.14** Training times–accuracy relationship curve for changing truncating probability

#### 4.4.4.2 Comparison of the Effects of EDA and GA

EDA and GA are adopted respectively for the model evolutionary computation. Their training effects are compared in the case of equal population. The model testing error rate after training is shown in Table 4.1. All the results are the average values obtained by 10 times of repeated computation.

It can be found in Table 4.1 that the error rate keeps about 10 % in the case of using GA and the changing parameters do not improve the results further. In addition, the error rate of the model evolving by means of GA shows a big variance,

**Table 4.1** Effects comparison of GA and EDA

| | EDA | | GA | |
|---|---|---|---|---|
| Population quantity | Selection probability | (Error rate, variance) | Population quantity | (Error rate, variance) |
| 100 | 0.1 | (0.0310,0.0404) | 0.08 | (0.1189, 0.0545) |
| | 0.2 | (0.0160, 0.0304) | 0.2 | (0.068,0.059) |
| | 0.3 | (0.0210, 0.0151) | 0.3 | (0.1220, 0.0594) |
| 300 | 0.1 | (0.0272, 0.0704) | 0.08 | (0.0994, 0.0652) |
| | 0.2 | (0.0050, 0.0087) | 0.2 | (0.0657, 0.0750) |
| | 0.3 | (0.0100, 0.0122) | 0.3 | (0.1153, 0.0501) |
| 1,000 | 0.1 | (0.0050, 0.0671) | 0.08 | (0.1036, 0.0503) |
| | 0.2 | (0.0090, 0.0104) | 0.2 | (0.1103, 0.0661) |
| | 0.3 | (0.0120, 0.0074) | 0.3 | (0.1153, 0.0501) |

and the training results are not stable enough with low overall performance. By contrast, with the same population quantity, the results obtained from training based on EDA can show an obvious lower error rate. The stability of algorithm can be improved a lot, which embodies fully the superiority of EDA in processing high-dimensional evolution. The experimental results indicate that EDA could better complete the parameters optimization of comprehensive fault judgment model for Maglev train with reasonable parameter setting.

#### 4.4.4.3 Performance Comparison of EDA with Other Machine Learning Algorithms

The C4.5 decision tree (C4.5Tree), Naïve Bayesian (Naïve Bayes), support vector machine (SVM), and neural network (neural network) are learning algorithms that are widely used in the field of machine learning, and they are capable of building the fault assessment model automatically based on training data. In order to compare the model optimizing performance of different machine learning algorithms for the assessment of Maglev train faults, WEKA software platform is adopted to conduct performance test of the learning algorithms. The parameters of every algorithm are set as follows:

1. C4.5 decision tree: confidence factor = 0.25, minNumObj = 1, numFolds = 3.
2. Naïve Bayes.
3. SVM: complexity parameter $C = 1.0$, use RBF = false.
4. BP neural network, learning rate = 0.2, momentum = 0.2, hidden layers = (attributes + classes)/2, training time = 500.
5. Fuzzy comprehensive judgment based on EDA evolution, the population quantity = 1,000, and the truncating probability is 0.3.

The performance of different algorithms is compared as shown in Table 4.2.

**Table 4.2** Comparison of EDA with other machine learning methods

| Learning algorithms | Testing method | Error rate (%) |
|---|---|---|
| J48(C4.5Tree) | 10-fold cross | 9.33 |
| | Test set | 12 |
| Naïve Bayes | 10-fold cross | 13 |
| | Test set | 16 |
| SVM | 10-fold cross | 8 |
| | Test set | 10 |
| Multilayer perceptron (BP neural network) | 10-fold cross | 6 |
| | Test set | 4 |
| EDA fuzzy comprehensive evaluation | 10-fold cross | 16.04 |
| | Test set | 1.00 |

It can be found in Table 4.2 that the 10-fold cross validation of EDA shows an inconspicuous effect which is even lower than that of other machine learning algorithms. However, it keeps the error rate within a very low range effectively in the process of validating the testing set, which precedes other algorithms obviously. These results are not contradictory for the following reasons. EDA is a learning method sensitive to the sample quantity, and its evolutionary effect can better do with the quantity of training samples. In particular, the larger the training individual dimensionality is, the more requirements of the sample are. However, the principle of 10-fold cross validation is to train by selecting 10 % of the training samples and test by using the rest 90 %, which means only 30 sets of the samples are used to train the individuals with $1 \times 33$ in dimension. Consequently, it is understandable that the training effect is not ideal. In the case of using the testing set for validation, however, if the amount of training samples reaches 300 sets, the assessment accuracy after training will increase by a large margin and meet the acceptance standard for application. Especially, after the evolution of EDA, the assessment accuracy can rise steadily with the continuous enriching of train operation data so that it can be qualified for ensuring the safe operation of Maglev train.

## 4.5 Conclusion

Firstly, in this section, the onboard control and diagnosis systems of Maglev train are introduced and presented, including their functional requirements and composition, data communication network, and hierarchy of onboard diagnosis system.

Secondly, the comprehensive assessment algorithm of onboard faults based on fuzzy comprehensive assessment is proposed. Then, the modeling principle for fuzzy comprehensive assessment of faults is discussed in detail, and the multilevel fuzzy comprehensive assessment model for Maglev train system faults is also built. In order to verify the proposed algorithm, the application and analysis of fuzzy comprehensive assessment of Maglev train system faults are made, including the framework of fuzzy comprehensive fault assessment process and experimental analysis.

Thirdly, in order to further assess the faults of Maglev train system, the comprehensive assessment method based on EDA is proposed. The basic theory of EDA is introduced and discussed in detail. The comprehensive assessment process of Maglev train faults based on EDA is presented, including modeling, encoding, calculation of fitness, building of probability estimation model, training, and testing model.

Lastly, the performance test and comparison are made, including the influence test of comprehensive judgment system parameters based on EDA, comparison of effects of EDA and GA, and performance comparison of EDA with other machine learning algorithms. The results show that the proposed comprehensive assessment method based on EDA can assess the faults of Maglev train efficiently and has more advantages than other methods.

## References


1. Chang Wensen. Development and control of Maglev train [C]. In: The 22th China control conference. Yichang; 2003, 8. p. 27–30
2. Sinha PK. Electronmagnetically suspension dynamics control [M]. London: Peter Peregrinus Ltd; 1987.
3. Chang Wensen. Maglev train: latest research trends [J]. Science. 1993;33(6):34–6.
4. Wu Xiangming. Maglev vehicle [M]. Shanghai: Shanghai Science and Technology Press; 2003.
5. Wang Lei, He Zhengyou. Technological characteristics of high-speed train communication network and its application [J]. Urban Mass Transit. 2008;11(2):57–64.
6. Lu Xiangyang, Zeng Rong, Liu Jun. On-board computer control network for "Central of China Star" EMUs [J]. Electr Drive Locomot. 2002;10(6):9–14.
7. Long Zhiqiang, Li Xiaolong, Zhou Wenwu. Study on the techniques of the locating and speed-measuring of the aerotrain [J]. J Natl Univ Def Technol. 2003;21(4):82–8.
8. Dai Chunhui, Xue Song, Long Zhiqiang. The signal disposal of position and speed detection sensors based upon long stators for Maglev train [J]. Chin J Sensors Actuators. 2009;22(6):822–6.
9. Zhou Jing, Wu Zhengkang. A fuzzy-based multi-hierarchy fault comprehensive evaluation model for EMU [J]. J Shanghai Tiedao Univ. 1999;20(2):45–51.
10. Long Zhiqiang, Lu Zhiguo. The fault diagnosis system of Maglev train based on fuzzy comprehensive evaluation [J]. Inf Control. 2004;33(2):227–30.
11. Long Zhiqiang, Lu Zhiguo, Huan Shang, Chang Wensen. Analysis and design in safeties and reliabilities of the suspension system of Maglev Train [C]. Proceeding of the 5th WCICA [C]. Hangzhou; 2004. p. 1819–23.
12. Long Zhiqiang, Cai Ying, Ye Jun. Analysis of the reliabilities of maglev train power system with DFTA method [C]. In: 19th international conference on Maglev and linear drivers. Dresden; 2006. p. 449–502.
13. Long Zhiqinag, Lu Zhiguo, Chang Wensen. Fault diagnosis to suspension system of maglev train based on fuzzy fault tree [J]. Control Decis. 2004;19(2):139–42.
14. Ho-Kyung Sung, Kim DS, Cho HJ, Yoo MH, Kim BS, Lee JM. Fault tolerant control of electromagnetic levitation system [C]. In: Proceedings of the Maglev 2004. Shanghai, 2004; p. 676–89
15. Hong-Ju Kim, Choon-Kyung Kim, Soonman Kwon. Design of a fault-tolerant levitation controller for magnetic levitation vehicle [C]. In: IEEE Proceeding of international conference of electrical machines and systems, Seoul, Korea; 2007, p. 1977–80.
16. Konstantinos Michail, Argyrios Zolotas, Roger Goodall, John Pearson. Fault tolerant control for EMS systems with sensor failure [C]. In: 17th Mediterranean conference on control & automation. Greece; 2009. p. 712–717.
17. Sung HK, Lee SH, Bien Z. Design and implementation of a fault tolerant controller for EMS systems [J]. Mechatronics. 2005;15:1253–72.
18. Long Zhiqing, Chen Huixing, Chang Wensen. Fault tolerant control on single suspension module of maglev train with electromagnet failure [J]. Control Theory Appl. 2007;24(6): 1033–7.
19. Zhiqiang Long, Song Xue, Zhizhou Zhang, Yunde Xie. A new strategy of active fault-tolerant control for suspension system of maglev train [C]. In: IEEE international conference on automation and logistics, Jinan, China; 2007, p. 88–93.
20. Xiong Jun. Genetic algorithm and its application [D]. PhD thesis. Nanjing: Nanjing University; 2004.
21. Zhou Shude, Sun Zengqi. A survey on estimation of distribution algorithms [J]. Acta Autom Sin. 2007;33(2):113–24.
22. Ren Yifeng. The estimation of distributions algorithm's analysis and its application [D]. PhD thesis. Shanghai: Shanghai Jiaotong University; 2005.

23. Wang Zhidong. The Modification and discussion of the estimation of distribution algorithm [D]. PhD thesis. Shanghai: Shanghai Jiaotong University; 2005.
24. Long Zhiqiang, Cai Ying, Xu Xin. Comprehensive fault evaluation on maglev train based on estimation of distribution algorithms [J]. Control Decis. 2009;24(4):551–6.
25. Sebag M, Ducoulombier A. Extending population based incremental learning to continuous search spaces [C]. In: Proceedings of the 5th conference on parallel problem solving from nature. Springer; 1998. p. 418–27.
26. MÄuhlenbein H. The equation for response to selection and its use for prediction [J]. Evol Comput. 1997;5(3):303–46.
27. Shapiro JL. Drift and scaling in estimation of distribution algorithms [J]. Evol Comput. 2005;13(1):99–123.
28. MÄuhlenbein H, Mahnig T. Convergence theory and applications of the factorized distribution algorithm [J]. J Comput Inf Technol. 1999;7(1):19–32.
29. Cristina G, Lozano JA, Larranaga P. Analyzing the PBIL algorithm by means of discrete dynamical systems [J]. Complex Syst. 2001;12(4):465–79.
30. Zhong Runtian. Distribution estimation algorithm and its application research [D]. PhD thesis. Hefei: University of Science and Technology of China; 2006.

# Chapter 5
# Maglev Train Control and Diagnosis Networks

In this chapter, the onboard control and diagnosis networks of Maglev train are presented in detail. The structures of control and diagnosis networks in Maglev train are firstly introduced. In order to effectively design and develop the onboard diagnosis network, and optimize the diagnosis and control of Maglev train, the communication performance of onboard diagnosis network is simulated and discussed. Secondly, according to the latest technology of high-speed train communication networks, a new integrated communication network including control and diagnosis networks based on the CANOpen technology is proposed. In addition, the framework and function of the integrated network are discussed and designed. Thirdly, ADS (autonomous decentralized system) and RoADS (role automation decentralization) ideas are introduced into the design of the diagnosis and control networks of Maglev train, which can better meet the demands of Maglev train diagnosis and control and improve the reliability, real-time, and autonomous property of diagnosis and control systems. In the end, aiming at the real-time demand of onboard communication platform in Maglev train, the design of communication platform based on RTLinux is realized, and the driver design for CPCI-CAN card is given in this chapter.

## 5.1 Introduction

Maglev (derived from magnetic levitation) is a system of transportation that can suspend, guide, and propel vehicles (predominantly trains) by using magnetic levitation from a very large number of magnets for lift and propulsion. In terms of the magnetism between magnets and magnetic components on the track, Maglev train is attracted to suspend above the track. The travel of train depends on the traction of linear motors in the vehicle. This method has the potential to be faster, quieter, and smoother than the wheeled mass transit systems. The power needed for



Z. Liu et al., *Maglev Trains*, Springer Tracts in Mechanical Engineering,
DOI 10.1007/978-3-662-45673-6_5

levitation is usually not a particularly large percentage of the overall consumption; most of the power used is needed to overcome air drag, as with other high-speed trains. Maglev train is one kind of high-speed passenger transport systems. Its average speed can arrive 400–450 km/h in the long distance travel [1].

The first commercial Maglev people mover was simply called "MAGLEV" and officially opened in 1984 near Birmingham, England. It operated on an elevated 600-m section of monorail track between Birmingham International Airport and Birmingham International railway station, running at speeds up to 42 km/h. The system was eventually closed in 1995 due to reliability problems. Perhaps the most well-known implementation of high-speed Maglev technology operating commercially is the Shanghai Maglev train, an IOS (initial operating segment) demonstration line of the German-built Transrapid train in Shanghai, China, which can transport people 30 km to the airport in just 7 min 20 s, achieving a top speed of 431 km/h, averaging 250 km/h [2].

Based on the suspension theory, there are two suspension methods, including electromagnetic and electrodynamic levitation [3–5].

For the electromagnetic levitation vehicle, through the electromagnetic field generated by electrical excitation at the track below the suspended electromagnets (or permanent magnets and excitation control coil), the electromagnet that is consisted of ferromagnetic components (steel track or long stator linear motor stator core) and the track attract each other. The train is lifted to suspend on the track, and the levitation gap between the electromagnet and the ferromagnetic track (called air gap) is generally about 8–10 mm. The train is driven through the linear motor traction, and the dynamic stability of the suspension air gap is ensured by controlling the excitation current of suspension magnet.

For the electrodynamic levitation vehicle, the train can only be levitated when the train reaches a certain speed. When the train travels, the moveable magnetic field (usually the low-temperature superconducting coil or permanent magnet) installed on the train generates the induced current in the suspended coil installed in the line. Their interaction can produce an upward magnetic force that makes the train suspend over a certain height of the rail (generally about 10–15 cm). The train's travel depends on the linear motor traction. Compared with the electromagnetic levitation, the electrodynamic levitation system cannot be levitated when the system is stationary. Only when the speed is over about 150 km/h, the system can be levitated. Since the suspension air gap of electrodynamic levitation system is large, the active controlling of suspension air gap is not necessary.

The onboard communication network in Maglev train mainly consists of two major parts, namely, onboard diagnosis network and onboard control network.

In the chapter, the basic knowledge and background about Maglev train will be introduced. Shanghai Line (Longyang Road – Pudong International Airport) will be taken as the example. The content includes the whole architecture, components, and key sections of Maglev train. In the onboard diagnosis network, there is a two-layer architecture including vehicle network (CAN bus) and train network (Ethernet) [6]. The vehicle network (CAN bus) is the core of onboard diagnosis network. Considering there are many electric and electronic devices in the vehicle, it is necessary to divide the CAN network into several parts. The diagnosed devices include all electric and electronic devices in the vehicle. The control network of Maglev train is composed of many hardwires that connect directly the controlled device. The control units and distribution unit of control orders are the key components in the onboard control network.

### 5.1.1 Diagnosis Network

The diagnosis network in Maglev train is one of most important systems involving the travel safety. During the travel of Maglev train, a great quantity of messages will be produced in the electric and electronic devices. In order to analyze and deal with the messages, it is very necessary to adopt the diagnosis system. The network structure is traditional supervision and control system frame, which includes two layers: the upper is Ethernet network structure and the lower is fieldbus network structure. The onboard diagnosis network is mainly used for the online diagnosis of whole train through the status acquisition of Maglev train's various components and some analog parameters to judge functional faults of onboard electrical and electronic components in Maglev train. The diagnosis network can also perform some unrelated safety control functions.

The main devices in diagnosis system include Vehicle Diagnosis Computer and its Display Module, Vehicle Operation Computer and its Display Module, Vehicle-Section Diagnosis Computer, two-layer Transmission Networks, and the Diagnosed Devices. The main functions of Vehicle-Section Diagnosis Computer mainly include reading repeatedly signals of Fieldbus Nodes, identifying the fault Fieldbus Nodes, producing diagnosis message based on judgment, and transmitting diagnosis message to Vehicle Diagnosis Computer.

The main functions of Vehicle Diagnosis Computer include receiving the fault messages from Vehicle-Section Diagnosis Computer, transmitting fault messages to Control Center, displaying fault messages, storing fault messages, receiving control signals and transmitting to Vehicle-Section Diagnosis Computer, identifying the fault of Vehicle-Section Diagnosis Computer, and transmitting and replicating data.

Considering that many electric and electronic devices exist in the vehicle, it is necessary to divide the CAN bus into several parts, which can not only improve the CAN bus network efficiency but also enhance the security of network. Thus, the fault of any bus part in CAN bus cannot result in the collapse of the whole network.

The diagnosed devices include all electric and electronic devices in the vehicle. There are Magnet Levitation Control Unit, Magnet Guidance Control Unit, Magnet Brake Control Unit, Linear Generator Converter, Energy Network Distributor, Energy Network Controller, Vehicle Control Unit, Door Controller, Location Measurement Unit, Air-Conditioning Controller, Inverter, Fan Unit, and so on.

The number of diagnosed devices is more than 100 in each section of vehicle. The structure of diagnosis system is shown in Fig. 5.1.

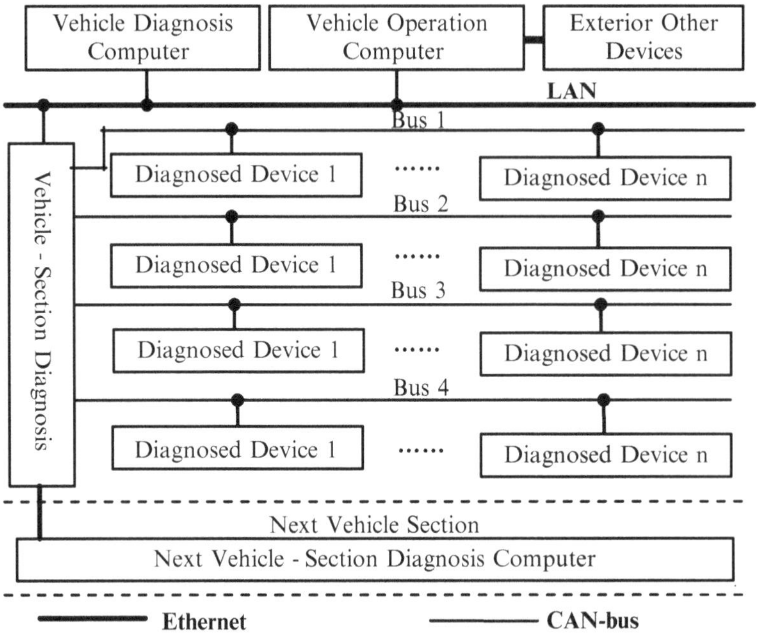


**Fig. 5.1** The frame of diagnosis system (Reprinted from Ref. [7], Copyright 2006 with permission from IEICE)

The message current model of diagnosis system is shown in Fig. 5.2. The fault judgment of diagnosed devices depends on the inner algorithm in Vehicle-Section Diagnosis Computer. Vehicle-Section Diagnosis Computer acts as the most important role in the diagnosis system [7].

### *5.1.2 Control Network*

The control system of Maglev train takes charge of the transmission and performance of control orders and safe status signals (supervisory signals). It is the "Nerve Centre" of train safe traveling. The control network in Maglev train is responsible for the control functions of electric and electronic devices. The main functions of control network are listed below.

On the one hand, the control orders from operation control centre on ground are sent, processed, and transmitted to the corresponding electric and electronic devices in Maglev train. The corresponding electric and electronic devices will carry out their functions, such as levitation, guidance, driving, and so on. On the other hand, the supervisory signals related with safety will be produced by the corresponding electric and electronic devices in Maglev train, such as levitation, guidance, driving, and so on.

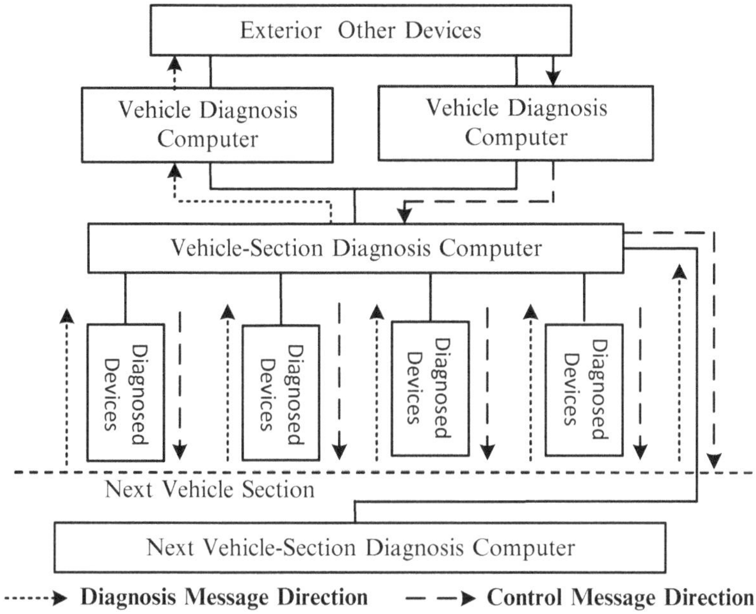


**Fig. 5.2** The message current model of diagnosis system (Reprinted from Ref. [7], Copyright 2006 with permission from IEICE)

In order to ensure the transmission of messages, the age-old hardware line network is adopted in the traditional Maglev train. In each vehicle, there are two Vehicle Control Units and one Vehicle Control Dispatch Units. All control signals are synchronously sent by two Operation Control Systems in double redundancy means. Then, in the two Vehicle Control Units, after the "OR" Boolean calculation, the control signals are transmitted and dispatched to the corresponding electric and electronic devices in double redundancy means.

All supervisory signals related with safety are transmitted to the two Vehicle Control Units in double redundancy means. In the two Vehicle Control Units, after the "AND" Boolean calculation, the supervisory signals are transmitted to Operation Control System [8]. That is to say, the transmission of control signals is bidirectional, while supervisory signals are unilateral. The Vehicle Control Units in middle vehicle not only supervise the status of electric and electronic devices in self-vehicle but also receive and transfer the supervisory signals from Vehicle Control Units of conjoint vehicles. The structure of control network in Maglev train is shown in Fig. 5.3.

In order to ensure the security and correctness of control orders during the course of transmission, the control network must add many hardware line connections between operation control system above and electric and electronic devices below. In Fig. 5.3, the hardware lines actually represent many hardware line bundles, which mean the complexity and redundancy of control network. In the control network of

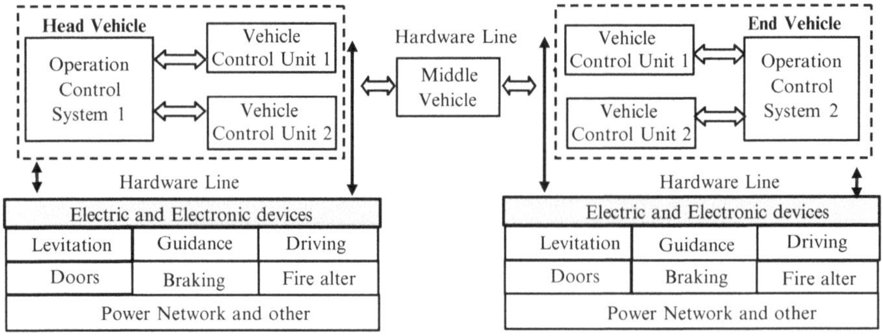


**Fig. 5.3** The structure of control network (Reprinted from Ref. [7], Copyright 2006 with permission from IEICE)

Maglev train, each control signal must be represented by voltage difference on two twist hardware lines. Considering the redundancy of control signals, corresponding spare hardware lines should be equipped in the control network, which results in very large number of hardware lines. In fact, there are about several hundreds of hardware lines for control signal transmission in each vehicle, though the type number of control signal is only more than eight.

## 5.2 Communication Simulation of Onboard Diagnosis Network

The onboard diagnosis network is the nervous system of Maglev train, which connects all controllers, sensors, and corresponding devices together to realize the information acquisition and control. In order to study the security and reliability of onboard diagnosis network, the simulation models of onboard diagnosis network of Maglev train with OPNET are built to analyze the network's performance, such as response error and bit error rate on the network load and throughput and node-state response. The online diagnosis of all onboard electric and electronic components in Maglev train can be carried out, and some control functions with unrelated safety can be completed through onboard diagnosis network. In one vehicle, a CAN communication network with four CAN buses includes vehicle diagnosis computer and all onboard electric and electronic components. Thus, the fault of any bus part in CAN network will not result in the collapse of whole network. Each CAN bus is composed of about 30 CAN nodes through which all onboard electric and electronic components can be connected together. In one train, the vehicle diagnosis computer, train diagnosis computer, and train operation computer constitute an industrial Ethernet, which is used to connect all vehicles in Maglev train.

In order to effectively design and develop the onboard diagnosis network, and optimize the diagnosis and control of Maglev train, it is very necessary to obtain the communication performance of onboard diagnosis network.

In the section, the simulation model of onboard diagnosis network is built with OPNET to analyze the network's performance, such as response error and bit error rate on the network load, throughput, and node-state response. The vehicle network (CAN bus) is the core of onboard diagnosis network. CAN is one of serial communication buses, which is proposed and developed by BOSCH, and has become an international communication standard. Many features of CAN, such as message priority, multi-master work mode, flexible set, high reliability, and real time, make it applicable to train communication. Compared with train communication network (TCN), Lonworks, and WordFIP, low price and flexible set are CAN's advantages [9]. Based on ISO11898 documents, CAN includes two-layer mode (data link layer (DLL) and physical layer). Data link layer includes logical link control (LLC) and medium access control (MAC). The receiver and filter, overload notice, and recover management of the frame are completed in LLC. MAC is the core of CAN protocol, in which transport protocols, such as control frame, arbitration, error check, fault confinement, and so on, are defined [10]. In Shanghai Maglev Line, CAN is adopted as the communication network in onboard diagnosis system of high-speed Maglev train [11]. In the onboard diagnosis system, the vehicle diagnosis computer is the core unit, which is in charge of the whole diagnosis network in one vehicle. These functions include sending network messages such as period messages, control messages, and query messages to four CAN buses, receiving diagnosis messages from all onboard electric and electronic devices of four CAN buses, and translating protocols between CAN and Ethernet [12].

## 5.2.1 *Modeling and Simulation of Diagnosis Network*

### 5.2.1.1 CAN Node Model

Based on the actual structure of diagnosis network, optimal network engineering tool (OPNET) is adopted to build the model of whole onboard diagnosis network [13]. The hierarchy modeling (network model, node model, and process model) is used to build the onboard diagnosis network model. The network model is the network topology structure including the distribution and number of nodes. The node model is composed of process models, which can constitute a complete protocol stack and reflect the features of actual devices. The process model is composed of finite-state machines (FSM), by which the complete protocols can be described. The CAN node model in OPENT is built in Fig. 5.4 [14].

CAN node model is mainly made up of application layer, data link layer, and physical layer. The application layer includes input, output, and "config" process modules, which is used to produce and respond messages. The functions

Fig. 5.4 CAN node model

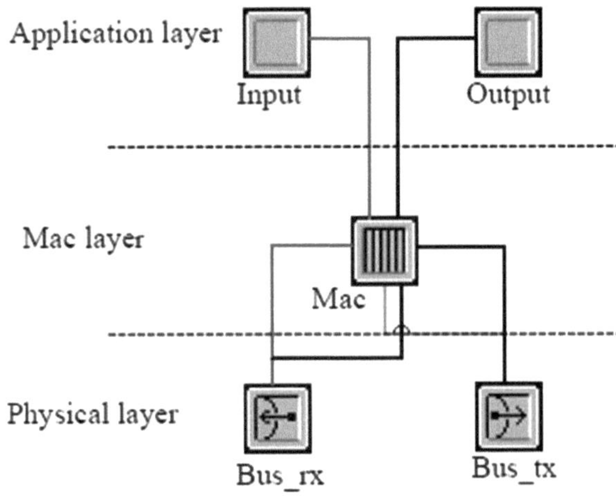


of application layer are realized through input and output modules, such as sending data messages produced in application layer to the bus or receiving data messages from the bus. The "config" process module is used to declare exterior files, by which the corresponding attributes of the whole network can be configured.

MAC is the most key part in CAN protocol, which processes the receiving messages from the bus and the sending messages from application layer, such as responding message frame, control frame, arbitration, checking error, and error definition.

For the model in Fig. 5.5, the system is firstly initiated to obtain the parameters of bus and directly enter into idle state. In idle state, the process can enter the sending or receiving state. The sending state means that the process can obtain data from application layer, seal data based on the data format of CAN protocol, send the data to physical layer, and return to the idle state. The receiving state means that if the messages arrive in bus, the receiving state is transferred. After receiving data, arbitration state is transferred. If new messages arrive in the next time interval, the system will return to receiving state. In the arbitration state, if no new messages arrive in a bit time, system will directly enter into decision state. The messages can be checked and processed in decision state. If the received messages are errors or there is no response, the system will enter into the state of receiving error and send error frame.

When the message receiving is successful, the system will collect error statistics. If the error number is more than 128, the system will enter into only sense state, or else enter into waiting state. In waiting state, if some messages are sent, the system will enter into the sending mode. If some messages are received, the system will enter into idle mode. During 8-bit time waiting in the only sense mode, if some messages are received, the system will enter into receiving state. If some messages are sent, the system will enter into sending state. If no data are sent or received, the system will enter into idle state directly.

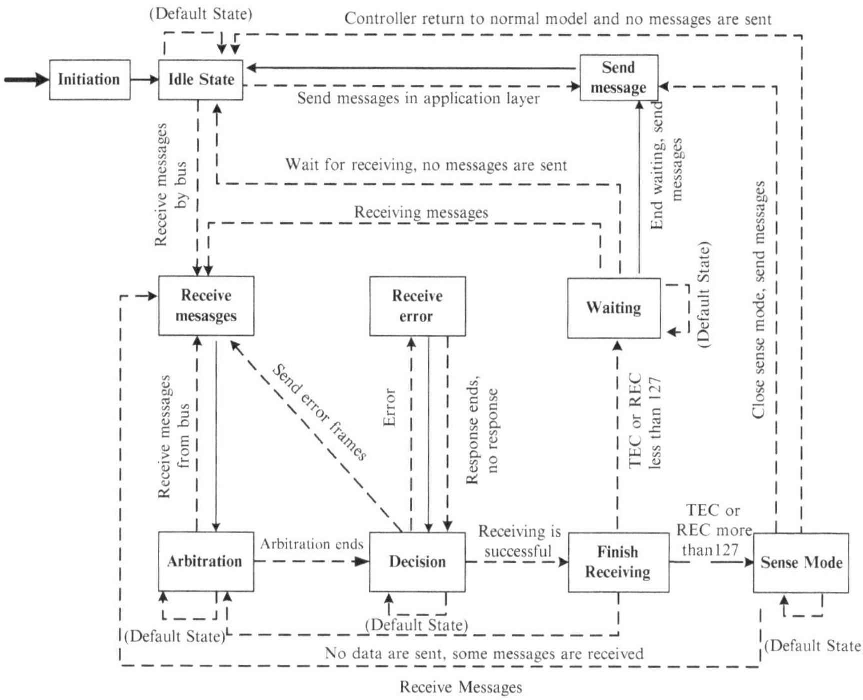


**Fig. 5.5** MAC structure of CAN node (Reprinted from Ref. [15], Copyright 2009 with permission from Journal of Southwest Jiaotong University)

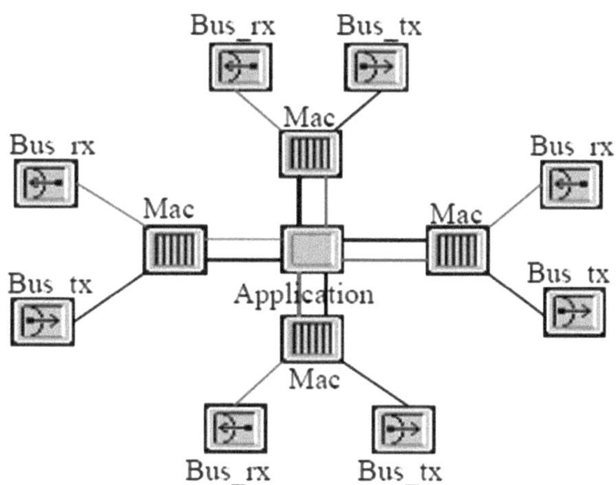


**Fig. 5.6** The model of vehicle diagnosis computer with OPNET (Reprinted from Ref. [15], Copyright 2009 with permission from Journal of Southwest Jiaotong University)

#### 5.2.1.2 CAN Network Model

The model of vehicle diagnosis computer with OPNET is shown in Fig. 5.6.

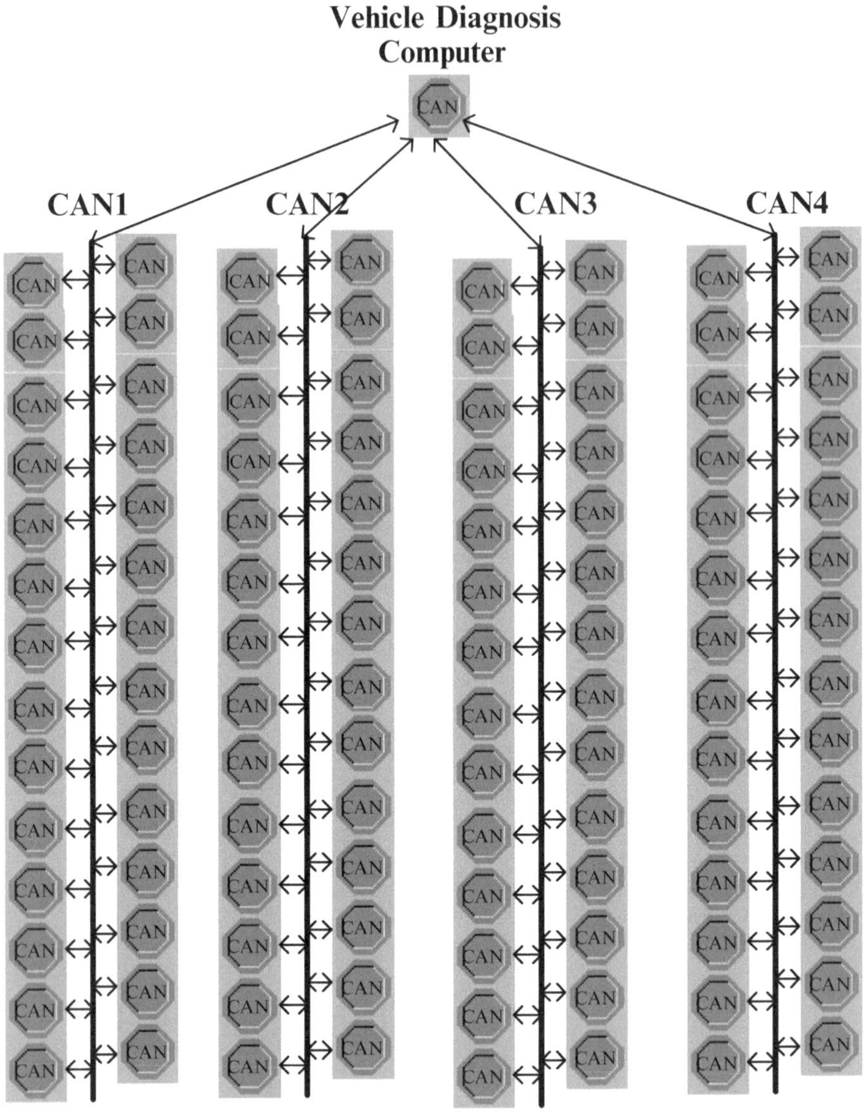


**Fig. 5.7** The model of vehicle diagnosis network with OPNET (Reprinted from Ref. [15], Copyright 2009 with permission from Journal of Southwest Jiaotong University)

In the model, the vehicle diagnosis computer includes four CAN buses as shown in Fig. 5.7. Each CAN node represents the actual corresponding electric or electronic device as shown in Fig. 5.8. The vehicle diagnosis computer means the main node of four CAN buses. The main node periodically sends the broadcast message in a standard frame to four CAN buses in each 1–1.5 s. Meanwhile, the

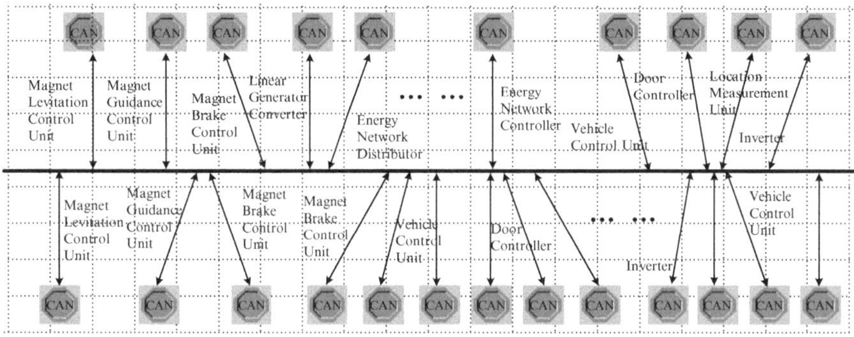


**Fig. 5.8** The actual model of single CAN bus with OPNET (Reprinted from Ref. [15], Copyright 2009 with permission from Journal of Southwest Jiaotong University)

**Table 5.1** The number and properties of each device node

| Device node | Number | Data |
|---|---|---|
| Magnet Levitation Control Unit | 6 | Uniform(0.02) |
| Magnet Guidance Control Unit | 6 | Uniform(0.02) |
| Magnet Brake Control Unit | 2 | Uniform(0.02) |
| Door Controller | 4 | Uniform(0.2) |
| Energy Network Controller | 4 | Exponential(0.1),uniform(0.01) |
| Passengers information | 2 | Exponential(1),uniform(0.1) |
| Fire alarm | 4 | Exponential(10),uniform(0.1) |

non-safe control messages, related with air conditioner switch, headlight switch, and so on, will be sent in a standard frame to four CAN buses.

When each node of electric or electronic device receives the broadcast message, the status message frame of each node will be sent to the main node as the response. The corresponding nodes associated with non-safe control can receive and process the control message based on the frame ID. If the main node cannot receive some node's status message, the main node will send the single-node inquiry message to the node. After three times, if the main node still cannot receive its status message, the node will be considered as the fault node.

Due to the fact that the control messages are periodically sent, the normal node associated with non-safe control can receive the control messages. According to the vehicle's length, the communication speed of 500 kbps is adopted in the model. In the single CAN bus, the number and properties of each device node are listed in Table 5.1, where CAN1 is taken as an example. "Uniform(m)" and "exponential(n)"mean the unvaried and average period with exponential distribution.

## *5.2.2 Simulation Results*

When the speed of Maglev train is 400 km/h, the train will move forward 11 m in each second. The real time and reliability of the onboard diagnosis system must fully be considered. According to normal network delay model, the application layer delay, waiting delay, sending delay, transmission delay, and receiving delay should be included. Considering that the application layer delay is related with the hardware performance and actual applications, it is not considered in simulation. In addition, the transmission delay is related with communication length, so the delay with short communication length can be ignored. The standard of communication network QoS (Quality of Service) mainly includes network delay, bus load, response time, network throughput, and so on. In the chapter, the main influencing factors of onboard diagnosis network's performance are considered. The simulation sections are listed below [15]:

1. The analysis of network performance for normal communication, such as the relationship between bit error rate, response error and bus load, throughput, response time, and error statistics.
2. The performance of onboard diagnosis network is analyzed when network topology is changed, such as changing the status of some electric and electronic device nodes, adding communication device node, and cutting communication device node from the network.
3. The network performance when changing the frequency of sending and receiving data packets for the device nodes is analyzed. In order to obtain high real time, the time interval of sending message for device node can be reduced.

The network performance is analyzed by changing the transfer characteristic of the device nodes. The relationship between BER (bit error rate) of onboard diagnosis network and bus load is shown in Fig. 5.9 as well as the relationship between BER and bus throughput in Fig. 5.10:

- When BER is 0, the bus load of onboard diagnosis network is about 11 %.
- When BER is 0.0001, the bus load of onboard diagnosis network is about 15 %.
- When BER increases 0.001, the bus load of onboard diagnosis network firstly increases sharply, then decreases with time, and keeps at about 16 % finally.

The analysis result is similar for the relationship between BER and bus throughput.

The relationship between frame frequency and bus load is shown in Fig. 5.11 as well as the relationship between frame sending frequency and bus throughput in Fig. 5.12. With the increase of frame sending frequency, the bus load and bus throughput will increase with time accordingly. The ETE (end to end) delay for different bus load and different priorities is shown in Fig. 5.13 as well as different BERs and different priorities in Fig. 5.14. When the bus load is 8 %, with different priority, the ETE delay basically keeps invariable at about 0.00018. When the bus load is 12 %, the curve is generally same. When the bus load is 31 %, the ETE delay will increase when the priority is 3 and then basically keep invariable at about 0.00024. The analysis result is similar to the relationship between BER and throughput.

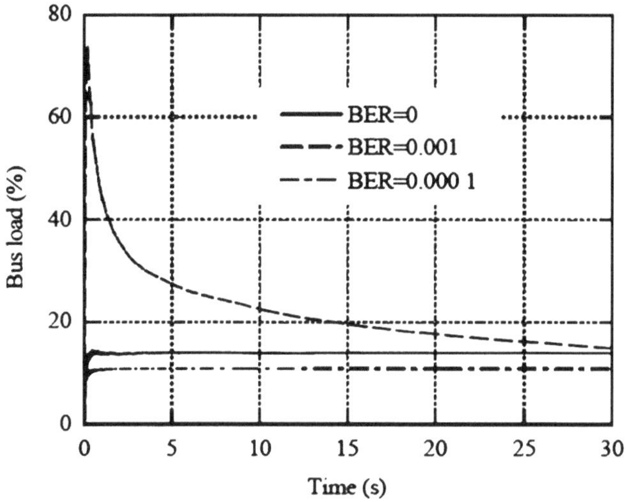


**Fig. 5.9** The relationship between BER and bus load (Reprinted from Ref. [15], Copyright 2009 with permission from Journal of Southwest Jiaotong University)

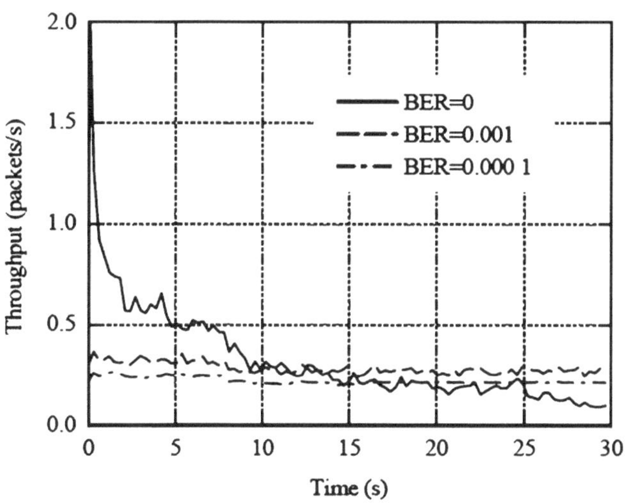


**Fig. 5.10** The relationship between BER and throughput (Reprinted from Ref. [15], Copyright 2009 with permission from Journal of Southwest Jiaotong University)

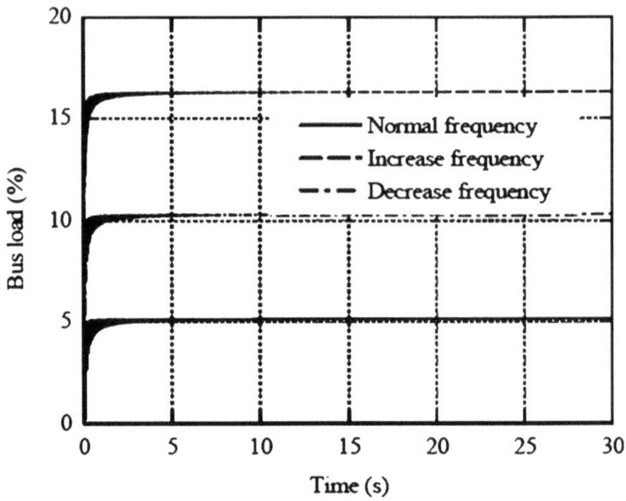


**Fig. 5.11** The relationship between frame frequency and bus load (Reprinted from Ref. [15], Copyright 2009 with permission from Journal of Southwest Jiaotong University)

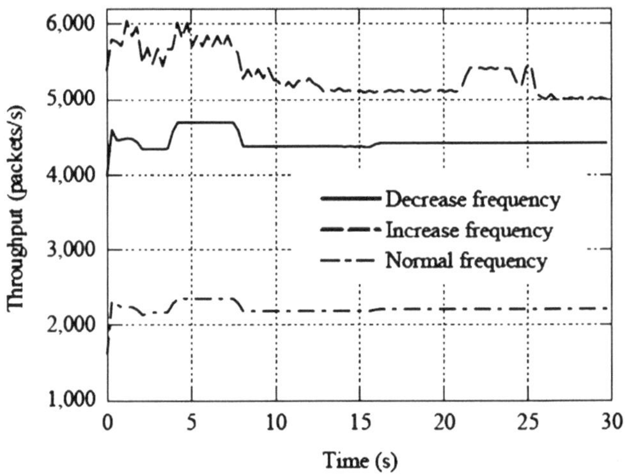


**Fig. 5.12** The relationship between frame frequency and throughput (Reprinted from Ref. [15], Copyright 2009 with permission from Journal of Southwest Jiaotong University)

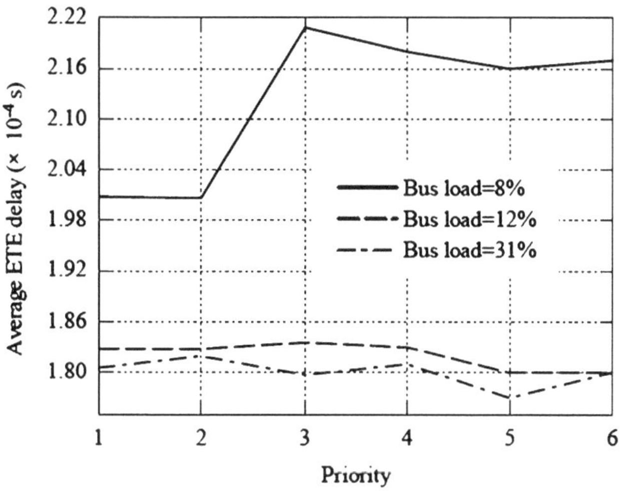


**Fig. 5.13** ETE delay for different loads and different priorities (Reprinted from Ref. [15], Copyright 2009 with permission from Journal of Southwest Jiaotong University)

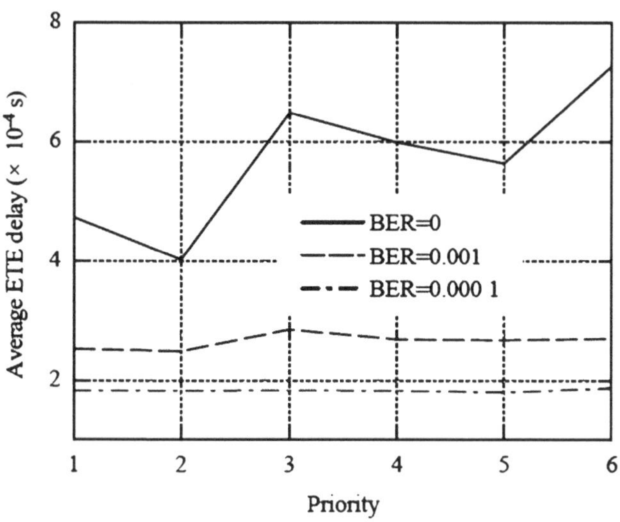


**Fig. 5.14** ETE delay for different BERs and different priorities (Reprinted from Ref. [15], Copyright 2009 with permission from Journal of Southwest Jiaotong University)

## 5.3 Integrated Network Design of the Control and Diagnosis Networks

### *5.3.1 Necessity of the Integrated Network for Control and Diagnosis Networks*

In Maglev train, the control and diagnosis networks are independently separated onboard, and they are different subsystems. The traditional hardware line communication is adopted in onboard control network. Two-level structure network including Ethernet and CAN is adopted in onboard diagnosis network. The main purpose of the structure is to make Maglev train system safer, since the control signals are closely related to the train safety, and diagnosis signals do not influence train safety. However, there are disadvantages as follows.

In the onboard diagnosis network, the application layer protocol of network node configuration is simple. The control network interconnection is poor. The large wiring causes the difficulty of construction and maintenance. The control network occupies a lot of vehicle space and increases the vehicle weight. The control network leads to a lot of reconnection problems and reduces the system scalability. It is very difficult to realize the interconnection and coordination between the control and diagnosis networks.

In fact, with the development of train communication network technologies, the traditional hardware line communication has been abandoned and fieldbus is generally adopted [16]. Some common train communication networks can be adopted to integrate the control and diagnosis networks of Maglev train. There are some advantages below:

1. In the onboard communication network of Maglev train, the intelligent devices are scattered to the field to perform a variety of sensing control alarm and computing functions, which can reduce the number of transmitter. The way can no longer need a separate regulator and calculation units, the traditional control systems, conversion, isolation unit, and its complex wiring.
2. The wiring is very simple in the onboard communication network of Maglev train. In general, a twisted pair or a cable can connect multiple devices; thus, the amount of cables, terminals, slot box, and the bridge is greatly reduced. In addition, the connection design and joints proofreading work is significantly reduced.
3. In the onboard communication network of Maglev train, the vehicle equipment has the ability of self-diagnosis and simple failure handling, which can produce the automatic diagnosis and maintenance information to deliver the control room. The maintenance personnel can easily check the operation status of all equipment, maintenance, and diagnostic information. The way can improve the failure analysis and exclusion rate and reduce the maintenance downtime and maintenance costs.

4. The intelligent vehicle equipment in the onboard communication network of Maglev train can reduce the error to a minimum during the course of transmission. The control accuracy only depends on the accuracy of the sensitivity and control accuracy of sensors and actuators, which can improve the system control accuracy greatly. Meanwhile, the internal functions of devices are strengthened due to the network structure simplification and fewer equipment connections, which can not only reduce the round-trip transmission of the signals but also improve the system reliability greatly.

## *5.3.2 Comparison of Mainstream Train Communication Networks*

### 5.3.2.1 MVB and WTB

MVB (Multifunction Vehicle Bus) is supported by an integrated bus controller, which can construct a simple device without the processor. MVB controller can realize the redundancy at the physical layer. Namely, a device through two mutual redundant can send the message from a line receiver and monitor another line at the same time. The special shielded twisted pair cable is used in WTB (Wire Train Bus). The layout of wires obeys the principle of redundancy, with one cable in each side of the vehicle. In order to adapt the frequent changes in train marshaling, WTB is designed to achieve the interconnection between vehicles through hand-plug jumper cable or auto-connector. WTB and MVB can divide data into process data (variable) and message data [17]. Depending on the two types of data, the second class architecture is defined.

### 5.3.2.2 LonWorks Bus

LonWorks, standardized by American National Standard ANSI/EIA709.1-1998, is a common control network protocol developed by American Echelon company. As one part of train communication protocol made by IEEE (type L), LonWorks with TCN constitutes IEEE1473. LonWorks is widely used in American train and is preliminarily used in China railway vehicles. The LonTalk protocol used by LonWorks realizes all seven layers of OSI reference model and provides a full range of network management services.

### 5.3.2.3 WorldFIP Bus

WorldFIP is one part of the European fieldbus standard EN50170 and is widely used in the field of energy, chemicals, transportation, and other industries as a common fieldbus standard. The WorldFIP is used in the A-GATE train control system developed by French company ALSTON, which is successfully used in TGV (train àgrande vitesse) trains. The WorldFIP bus adopts 3-layer structure and supplies the variable service, message service, and network management service.

#### 5.3.2.4 CAN Bus and Its Application Layer Protocol CANOpen

CAN bus is originally developed by German company Bosch for vehicle monitoring and applications in controlling data communication of electronic system [18]. Now it becomes the international standard ISO11898 (high-speed applications) and ISO11519 (low-speed applications) and accesses to a very wide range of applications, including train communication. However, being different from other fieldbus structures, CAN bus can only define the physical layer and data link layer, but the other layers are not specified. This design enables CAN to be widely adapted to variable application conditions, but inevitably brings some inconvenience to the user. Therefore, some European companies jointly developed the CAL (CAN application layer) protocol. Considering that CAL is just a pure application layer protocol and each user must design a sub-protocol in the process of usage, it is difficult to get uniform and wide range of applications. Then, based on CAL, Bosch developed a sub-protocol prototype, which has evolved into CANOpen protocol. CANOpen is designed with object-oriented idea and wins relatively wide application in train communication network of vehicle level due to its good modular characteristics and high adaptability [19].

China railways such as the Beijing Airport Line light rail and the second phase of Changchun light rail use CANOpen as the fixed connected unit vehicle-level communication network without reconnection application. The control signals and status information are transmitted via the network.

As shown in Fig. 5.15, Beijing Airport Line light rail adopts the fixed connected four cars grouping, with CANOpen as the vehicle-level communication network throughout the grouping unit to realize data communications. Since the grouping unit does not require the reconnection, the train-level communication network is absent. However, in order to meet the reconnection application, the variable expansions of train bus can be realized with different gateways. The entire CANOpen vehicle-level network is a redundant network consisting of two independent CAN buses [20]. Two VTCUs (Vehicle Train Control Unit) located in cabs in both ends of the train could manage the network as the masters of redundant network, while only one VTCU can control and manage the train network. There are two mutually redundant monitoring systems, collecting all kinds of status information of vehicle grouping unit through CANOpen network. All information is displayed simultaneously in two HMI (Human Machine Interface) screens of two VTCUs, and the screens are connected to VTCUs via RS232 interface. The rate of data transmission in CANOpen vehicle network is determined by the length of the CAN bus; therefore, the cable length of vehicle network requires the strict calculation, while CAN repeaters installed in the bus can split the bus length and improve

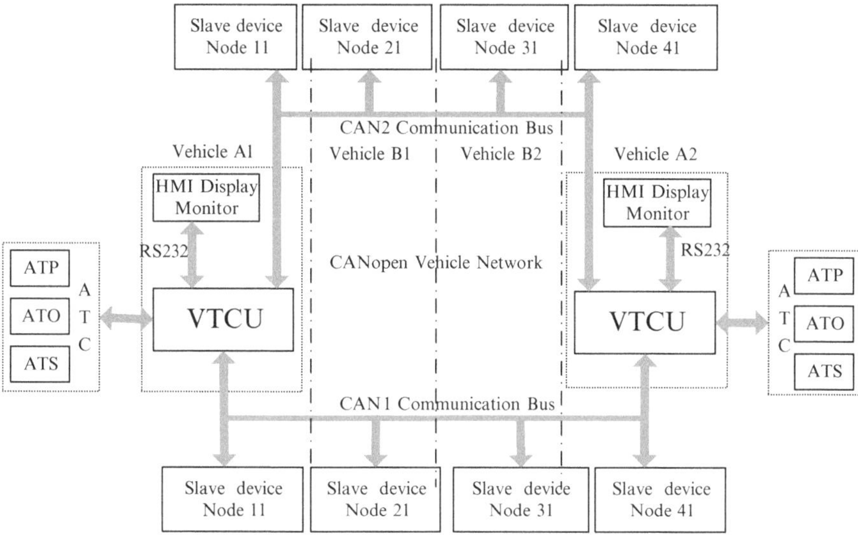


**Fig. 5.15** Beijing airport line light rail vehicle control and diagnostic system

the transmission rate. Distributed controllers of the slave devices can control the subsystems and collect status information through I/O.

The most essential status information and diagnostic information are transmitted over the network to VTCU which is connected to the display, and the status and diagnostic information will be shown in the display. The most essential control data are sent over the network to the controllers of each subsystem, which will achieve their control through digital or analog outputs with relay or thyristor. The VTCU and various subsystem controllers can be connected via the RS232 interface to a personal computer in which the maintenance, commissioning, and fault data analysis work are finished through the expert software.

### 5.3.3 *Architecture of System Bus of Vehicle Unit Based on CANOpen*

The system bus architecture of the vehicle marshaling unit consists of CANOpen vehicle bus which connects to all control devices of the entire marshaling units and realizes data transmission and internal logic control bus inside the subsystem devices which connects to I/O and distributed controllers.

As shown in Fig. 5.16, the node devices (including master and slave nodes) are connected to each other through CANOpen vehicle bus. Inside node devices, DC/DC converters supply the power for CAN transceivers (+5VDC). CAN transceivers are connected to CAN controllers (CAN chip) via optocouplers.

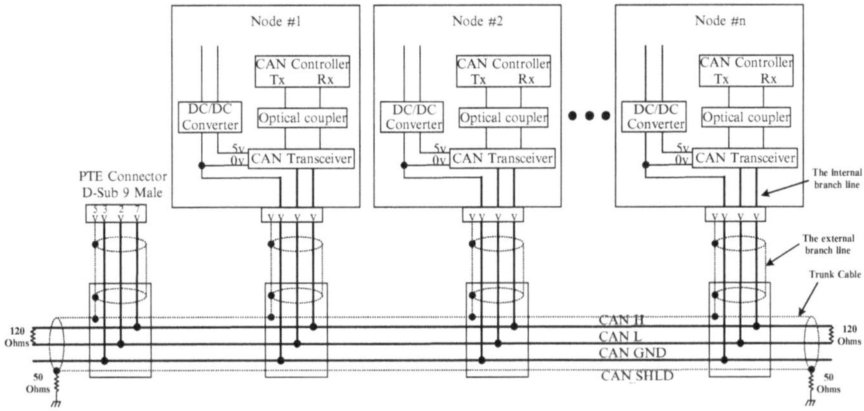


**Fig. 5.16** Connection of CANOpen vehicle bus devices

Several portable test equipment connectors are linked to the bus for the purpose of convenient connection of test equipment and complete the testing and debugging of bus communication. Two 120 Ω termination resistors are integrated into both ends of bus trunk cable, and the CAN cable shield is grounded through a 50 Ω resistor. The CAN cable binding node devices to the bus trunk is called a branch line whose length is specified according to network design. The internal logical control bus of node devices is used to transmit logical signals between modules such as controllers and I/O. Its logic supply is +5 VDC. The controller realizes the data collection and output through internal logical control bus and exchanges data with master node controllers or other node controllers through the vehicle bus. Selectron company CPU73X and CPU72X series PLC (programmable logic controller) is used in the Beijing Airport Line light rail network control system. Different I/O modules are chosen according to the different applications of node devices. For example, the airport line light rail air-conditioning control system mainly uses $8 \times 24$ VDC digital input module, $8 \times 24$ VDC digital output module with delay, and $4 \times 2$ mA Pt100/Pt1000 temperature collection analog input module.

Taking Beijing Airport Line light rail air-conditioning control system as example, the distributed controller CPU723 has two independent CAN interfaces. One is connected to the network node module DDC712 for I/O expansion, and the other is connected to the vehicle bus and communication as child node. CPU723 and DDC712 collect the digital input/output and analog input of the temperature sensor by the internal logic control bus, which is completely independent of the CAN bus. CANOpen devices model for transmitting and receiving frames are shown in Fig. 5.17.

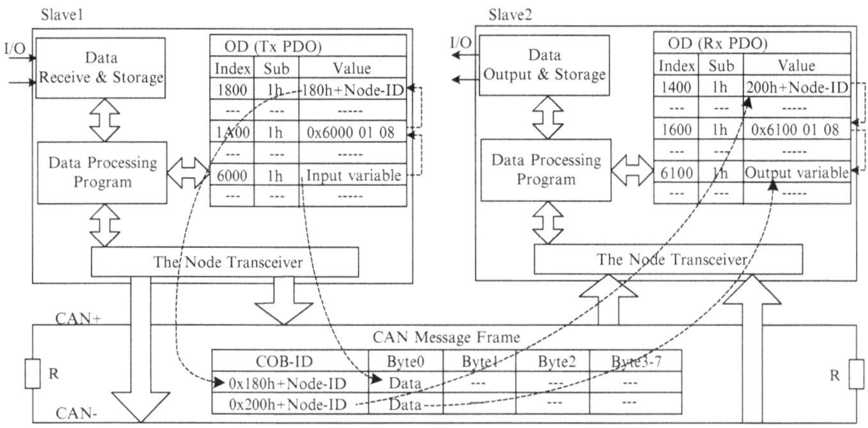


**Fig. 5.17** CANOpen devices model for transmitting and receiving frames

## 5.3.4 *CANOpen Vehicle-Level Network Devices Classification*

As the same as other vehicle buses, CANOpen vehicle bus also exists as the communication medium for transmitting and exchanging data between different devices [21]. Although the node devices on the bus vary in function and volume, all of them can exchange information through CANOpen vehicle bus and form a complete communication network. Lower layer protocol determines that the CANOpen bus has a linear form, and all devices that require the bus communication are connected to it as a node of the bus. Nodes on the bus are divided into master nodes and slave nodes according to their functions in network, while they can also be divided into the controller nodes and pure network nodes for I/O extension.

### 5.3.4.1 Master/Slave Devices and the Analysis of Their Functions

In CANOpen network, the master node works as the management device of the entire network. Since CANOpen data transmission applies to the competition instead of polling mechanism, the main function of master device is configuring parameters and managing the communication of all slave devices on the bus. When a slave device participates in the network communication, the master device can configure the communication parameters and manages the communication process of the slave device. The structure of master/slave devices and their functions are shown in Fig. 5.18.

There is no difference in hardware structure between master and slave devices from the data communication, that is to say, their processor modules can both exchange data with memories or I/O by the internal bus and communicate with other devices on the bus through transceiver. The configuration and management of the

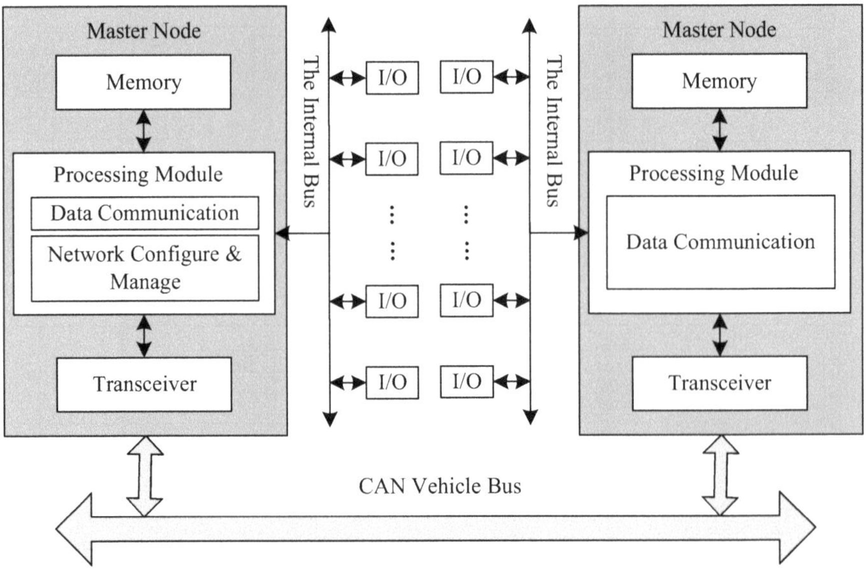


**Fig. 5.18** The structure and function of CANOpen master/slave devices

entire network are realized through software. In the process of booting the network, master devices verify the slave devices according to default network model in the program. If the slave device exists, its parameters are configured. Then, after the slave is successfully configured, NMT (Networked Media Tank) command is sent to start the slave that will enter normal operating status.

#### 5.3.4.2 Controllers, Pure Network Devices, and Their Functions Analysis

The changes from the centralized control to distributed control exist in vehicle control system. Doors and air conditioning, as well as the traction power supply system, braking system, passenger message display system, broadcasting system, smoke sensor, and other equipments, are dispersedly controlled by the independent controllers.

A controller, as a node in CANOpen bus, can communicate the status information, diagnostic information, and control information with train master control unit and other node devices. These node devices are called the controller nodes. A distributed controller can realize the system control and status monitoring by I/O modules connected to the internal bus. Sometimes, due to the limited quantity of I/O modules, the available interface cannot meet the application demands of certain subsystems.

One or more additional node devices used to count I/O modules are needed to add and collect remaining I/O data, which will be exchanged with system controller nodes by CANOpen network to achieve the goal of monitoring and controlling. These node devices used to extend the I/O interface are called the pure network nodes, which do not have any control behavior in the system according to the software program and only exist as the media for I/O data collection. There are two common modes in CANOpen vehicle-level communication, point-to-point master/slave communication mode and producer/consumer intelligent broadcasting mode. Point-to-point master/slave mode is a reliable basic communication mode, while broadcasting mode improves the communication efficiency and optimizes the network performance.

## 5.3.5 *Node Device Description and Performance Analysis*

### 5.3.5.1 Node Device Description in CANOpen Network

Strong Master Processing Unit
Network master node is responsible for the control and supervision of the entire network, communication between the display unit and radio system, and critical control and fault diagnosis on train.

Weak Master Processing Unit
Redundant equipment as the strong master node in the primary node can replace the master node when it is down or failure.

Key Control Unit
It can receive the control command of the main processing unit, perform the critical control of the vehicle, and transmit the security-related diagnostic data back to the main processing unit.

Step-Up Chopper Unit
It can collect and analyze the diagnosis signals of step-up chopper and send the computed data to the main processing unit. Some unsafe-related control can be performed.

Magnet Guidance Control Unit
It can collect and analyze the diagnosis signals of Magnet Guidance Control Unit, perform magnet guidance control commands, and send the computed data to the main processing unit.

Magnet Brake Control Unit
It can collect and analyze the diagnosis signals of Magnet Brake Control Unit, perform magnet brake control commands, and send the computed data to the main processing unit.

Magnet Levitation Control Unit
It can collect and analyze the diagnosis signals of Magnet Levitation Control Unit, perform magnet levitation control commands, and send the computed data to the main processing unit.

Inverter Unit
It can collect and analyze the diagnosis signals of inverter unit and send the computed data to the main processing unit. Some unsafe-related control can be performed.

Energy Network Distributor Unit
It can collect and analyze the diagnosis signals of Energy Network Distributor Unit and send the computed data to the main processing unit. Some unsafe-related control can be performed.

Energy Network Controller
It can receive the control command of the main processing unit, perform the energy network control commands, and transmit the security-related diagnostic data back to the main processing unit.

Door Control Unit
It can collect and analyze the diagnosis signals of door control unit and send the computed data to the main processing unit. Some unsafe-related control can be performed.

Location Measurement Unit
It can collect and analyze the diagnosis signals of Location Measurement Unit and send the computed data to the main processing unit. Some unsafe-related control can be performed.

Air-Conditioning Control Unit
It can collect and analyze the diagnosis signals of Air-Conditioning Control Unit and send the computed data to the main processing unit. Some unsafe-related control can be performed.

In the CANOpen network, the main processor uses 4 independent CAN ports to form four CAN buses. All diagnosed and controlled electric and electronic components are evenly distributed to four CAN buses. As the safety controller for the whole train, Key Control Unit is redundant. In each section of the vehicle, there are two redundant Key Control Units connected in different CAN buses.

In the whole control and diagnosis network based on CANOpen, there are four CAN buses. For three vehicle grouping, there are 91 nodes, respectively, 22 nodes in CAN 1, 21 nodes in CAN 2, 27 nodes in CAN 3, and 21 nodes in CAN 4. Through the competition, the packets can obtain the bus control in the CANOpen network. The priority of communication objects that send packets are directly related to the type and size of the node number. When the node number is assigned, according to the rule from high to low importance of equipment in the network, the node numbers from small to large are dispatched. Considering Key Control Unit is the critical component in the CANOpen network, its importance is highest except master processing units. The node number can be assigned from 6 to 8. Sequentially, other nodes' number can be assigned.

#### 5.3.5.2 Real-Time Analysis of the Communications Network

Data transmission delay model of CANOpen network is constructed below. The underlying protocol of CANOpen network is CAN. The delay of packet transmission in CAN bus has the significant influence on the whole network performance, which largely decides the real-time property of CANOpen network. Through the data transmission delay model, the real-time property of CANOpen network is analyzed below.

In this analysis, the reasonableness of the CANOpen network communication objects arranged in the actual project is explored and tested, which can avoid possible bus block and further optimize the network performance.

The packet transmission delay model $T_{delay}$ in CAN is defined as the time from sending packets of source node $T_s$ to receiving packets of destination node $T_d$, which includes three parts (total sending delay $T_{total}$ of source node, transmission delay $T_{trans}$ in network channels, and receiving delay $T_{rec}$ of destination node). The total sending delay $T_{total}$ includes the waiting time $T_{wait}$ and sending delay $T_{frame}$ of packets (transmission of frames). The waiting time $T_{wait}$ is the sum of queuing time $T_{queue}$ and block time $T_{block}$ [22]. The packet transmission delay model is shown in Fig. 5.19.

The queuing time $T_{queue}$ is the time from waiting for the data buffer of source node in the front of the packets to the packets sending completion. $T_{queue}$ depends on the block time $T_{block}$ of packets in front of the queue, the packet length, and the sending time $T_{bit}$ of each bit. In the switched Ethernet, $T_{queue}$ is the determinant factor of transmission delay. In CANOpen network, considering that there is the minimum inhibitory time limit of sending real-time data and old packets are often discarded, the queuing time $T_{queue}$ is considered as zero in general.

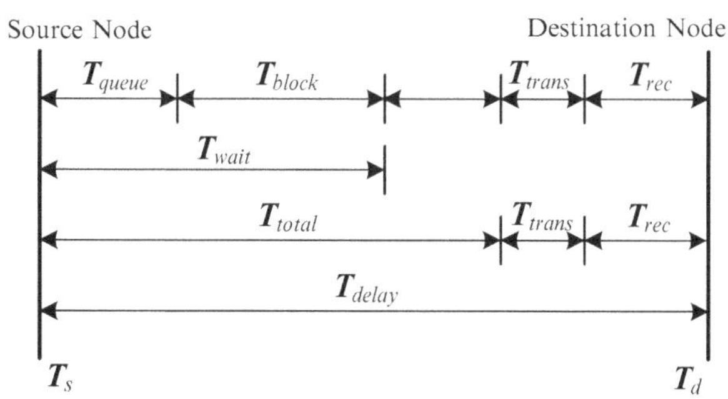


**Fig. 5.19** The packet transmission delay model

The block time $T_{\text{block}}$ is the time from starting to send a packet to get the bus control. In CANOpen network, $T_{\text{block}}$ mainly depends on the by-bit ruling access mechanisms of CSMA/CD (Carrier Sense Multiple Access/Collision Detection) [23]. It includes the waiting time for other nodes' sending higher-priority packets and the waiting time for sending or resending packets when conflict or error occurs.

According to the determinate bus transfer rate, the sending delay $T_{\text{frame}}$ is the time of packets being sent to the network, after the node obtaining the bus control. It depends on the total frame length and sending time $T_{\text{bit}}$ of each bit.

The transmission delay $T_{\text{trans}}$ is the transmission time from source node to destination node in CANOpen network. The receiving delay $T_{\text{rec}}$ is completely receiving packet time of destination node. The data transmission delay model of CANOpen network is shown as follows:

$$T_{\text{delay}} = T_{\text{d}} - T_{\text{s}} = T_{\text{total}} + T_{\text{trans}} + T_{\text{rec}} = T_{\text{queue}} + T_{\text{block}} + T_{\text{frame}} + T_{\text{trans}} + T_{\text{rec}} \tag{5.1}$$

As described above, since the real-time and safety requirements are very high for the vehicle communication network applications, there are not old packets in general. Hence, $T_{\text{queue}}$ is zero. The transmission time $T_{\text{trans}}$ is ignored because the transmission time between the two nodes is too small. Therefore, $T_{\text{trans}}$ is zero. Meanwhile, the data acquisition of receiving node from bus is immediate; hence, $T_{\text{rec}}$ is zero in general. The simplification mathematical model of data transmission delay for CANOpen network can be obtained as follows:

$$T_{\text{delay}} \approx T_{\text{block}} + T_{\text{frame}} \tag{5.2}$$

Taking 11 identifier of the standard frame as an example, the frame time of CAN bus depends on the frame length and transmission time of each bit. The frame length $L_{\text{frame}}$ is composed of frame format bit $N_{\text{frame}}$, data bit $N_{\text{data}}$, and bit stuff bit $N_{\text{stuff}}$. Corresponding mathematical form is shown as follows:

$$T_{\text{frame}} = L_{\text{frame}} \times T_{\text{bit}} = (N_{\text{frame}} + N_{\text{data}} + N_{\text{stuff}}) \times T_{\text{bit}} \tag{5.3}$$

The frame format bit $N_{\text{frame}}$ is composed of frame start, arbitration field, control field, check field, response field, and frame end. For the standard frame, $N_{\text{frame}} = 44$ bit. The data bit $N_{\text{data}}$ is the actual bit number in data fields of frame, and 0 bit $< N_{\text{data}} <$ 64 bits (8 bytes). Considering that the bit filling code mechanism is adopted in CAN bus, when 5 continuous polarity bit streams need to be sent, a complement bit must be automatically inserted at the back of actual 5 continuous polarity bit streams. $N_{\text{stuff}}$ is the number of inserted complement bit.

The bit format of each packet is invariable, and the data bit and filling bit is variable. Based on Formula (5.3), the frame sending time can be computed for the different data length and filling conditions when the bus transmission speed is 250 kbps in CANOpen network. The results are listed in Table 5.2. It can be

**Table 5.2** Frame sending time for different conditions

| Data length/byte | 0 | 1 | 2 | 3 | 4 | 5 | 6 | 7 | 8 |
|---|---|---|---|---|---|---|---|---|---|
| No filling bit sending time/ms | 0.176 | 0.208 | 0.240 | 0.272 | 0.304 | 0.336 | 0.368 | 0.400 | 0.432 |
| Max filling bit sending time/ms | 0.192 | 0.228 | 0.268 | 0.304 | 0.344 | 0.380 | 0.420 | 0.460 | 0.496 |

found that the sending time of a frame is influenced by the data length and filling bit number for same data length. The sending time difference is about 8.5–13.2 %.

$T_{\text{block}}$ is mainly influenced by the occupied bus time of frame space because of other nodes' sending higher priority packets. Because of the conflict or error for packets, the relatively resending probability is less. Hence, when the bus transmission is normal and the waiting time is limited, $T_{\text{block}}$ can be defined as follows, where $k$ is the priority number and $k = 1, 2, 3 \ldots$. $T^k_{\text{block}}$ is the block time of $k$ priority. $T_{\text{sent}}$ is the remaining time when the current node of control bus is sending packets. $T^k_{\text{rblock}}$ is the block time when all other nodes are waiting to send packets whose priority is higher than that of $k$ node. $T^k_{\text{rblock}}$ can be represented as follows:

$$T^k_{\text{block}} = T_{\text{sent}} + T^k_{\text{rblock}} + T^k_{\text{Random}}, k \geq 2 \tag{5.4}$$

$$T^k_{\text{rblock}} = T^{k-1}_{\text{rblock}} + T^{k-1}_{\text{frame}} + 3T_{\text{bit}} \Rightarrow T^k_{\text{rblock}} = \sum_{j=1}^{k-1} \left(T^{k-1}_{\text{frame}} + 3T_{\text{bit}}\right), T^k_{\text{rblock}} = 0, k = 1 \tag{5.5}$$

$T^k_{\text{Random}}$ is the occupied bus time of sending packets whose priority is higher than $k$ priority, when the node with $k$ priority is waiting to send packets. $T^k_{\text{Random}}$ has certain degree of randomness.

Hence, when $k = 1$, namely, the priority $T^k_{\text{Random}}$ of the packets that is waiting to be sent is zero, the final expression of $T_{\text{delay}}$ is shown as follows:

$$T^k_{\text{delay}} = T_{\text{sent}} + \sum_{j=1}^{k-1} \left(T^{k-j}_{\text{frame}} + 3T_{\text{bit}}\right) + T^k_{\text{Random}} + T^k_{\text{frame}}, k \geq 2 \tag{5.6}$$

$$T^k_{\text{delay}} = T_{\text{sent}} + T^k_{\text{frame}}, k = 1 \tag{5.7}$$

It can be found that the main influence factors of packet transmission delay include the level of priority $k$, the data bit length of all packet frames in bus, the filling bit number, and the number of newly produced frames with high priority. The priority $k$ of packet frame almost dominates the frame transmission delay time. With the increase of $k$, which means the decrease of priority, the packet transmission delay will monotonically increase. The packet frame with the highest priority in network has the block and delay problems in the transmission process. The block time is $T_{\text{sent}}$.

## 5.4 Diagnosis Network Based on ADS

As the breakthrough over the conventional systems, ADS has been proposed and its technologies have been developed. The decentralized system is defined as such a living thing that is composed of largely autonomous and decentralized subsystems. Each autonomous system in autonomous decentralized system that is not controlled by other subsystems or systems is mutually independent and completes its own function. Meanwhile, these subsystems can initiatively transmit the internal information to external subsystem [24].

In virtue of the theory of ADS, some basic ideas of ADS are applied in new diagnosis system. The structure, component parts, and diagnosis method of new diagnosis system are proposed, designed, and discussed in detail. With the in-depth study about ADS and its corresponding theory, the traditional automation systems gradually have more demands for their applications and development. The space time dynamic evolution representation model of system potential functions corresponds to a kind of special automation system technology. The technology is called role automation decentralized system (RoADS). Considering many disadvantages of the traditional control and diagnosis networks, some RoADS ideas will be adopted in the new control and diagnosis network of Maglev train. In addition, their design principle and plan based on RoADS will be proposed, discussed, and analyzed in the section.

### *5.4.1 Design of Diagnosis System in Maglev Train*

#### 5.4.1.1 Defects' Discussion of Diagnosis System in Maglev Train

Considering the convenience of discussion, the diagnosis system frame of Maglev train is divided into 4 layers: Device Node Layer, Fieldbus Layer, Diagnosis Layer, and Decision Layer. The defects of the diagnosis method are listed below.

Device Node Layer
Each device independently sends status parameters by itself. There are no relationships among other electric and electronic devices, which mean that all status parameters of electric and electronic devices must be sent to CAN bus network. In fact, it is not necessary to send status parameters if status parameters of other electric and electronic devices are learned.

Fieldbus Layer
It is not proper to simply divide fieldbus into four CAN buses. These corresponding factors should be thought about including the devices relativity in the same bus, the number of bus, and so on.

Diagnosis Layer
In Diagnosis Layer, analyzing, synthesizing, processing fault messages from diagnosed devices, and transmitting fault information are, respectively, completed. But processing some fault information can be realized in Device Node Layer instead of Diagnosis Layer, such as the diagnosis of single electric and electronic device.

Decision Layer
The functions in Decision Layer consist of producing, displaying, storing, transmission of fault information, and diagnosing of the whole vehicle fault. There are some functions overlapping with Diagnosis Layer.

In order to meet the demands including flexibility and coordinability of the whole system, online diagnosis of electric and electronic devices, real-time property of fault information transmission, correctness of fault message, and reliability of data communication, it is very necessary to adopt new diagnosis system. The basic ideas of ADS will be simply introduced below.

#### 5.4.1.2 The Ideas of ADS

ADS has been proposed to resolve the online property of online expansion, online maintenance, and fault tolerance in a system, which means that the system can continue operation during partial expansion, maintenance, and at the time of a partial fault. The ADS is defined as the characteristics that each subsystem can control itself and coordinate with all of the other operating subsystems except the inoperable subsystems. In this system, the autonomous systems have two properties: autonomous controllability and autonomous coordinability [25].

**Autonomous Controllability** If any subsystem fails, the other subsystems can continue to manage themselves.

**Autonomous Coordinability** If any subsystem fails, the other subsystems can coordinate their individual objectives among themselves.

These two properties assure online expansion, fault tolerance, and online maintenance of the system.

There are four levels in ADS architecture:

**LN (Logical Node)** LN is the most functional unit in the autonomous decentralized system. Within a data field, a number which uniquely identifies a Logical Node is called a Logical Node Number (LNN).

**DF (Data Field)** DF is the space of transmitting messages in the autonomous decentralized system. In the physical concept, DF corresponds to the transmission network or memory.

**MCG (Multi-cast Group)** A multi-cast group refers to a group of nodes set up within a data field. In order to make the transmission and receipt of messages more convenient, the Logical Nodes of which logical connection is similar are divided into a broadcast group.

**Domain** A domain referring to an upper concept of data fields is a group of data fields. Generally each domain is defined as a local site and connected with other domains through wide area communication networks. The concept of Domain corresponds to LAN (local area network).

The communication protocol (called autonomous decentralized protocol) is based on the concept of autonomous decentralized. ADP (autonomous decentralized protocol) whose lower layer is TCP, UDP/IP, or Ethernet is considered as the application development on the application layer of TCP.

A new diagnosis system of Maglev train based on some basic ADS' ideas and its similar architecture will be designed. According to the ADS' ideas, its construction frame can be drawn in Fig. 5.20.

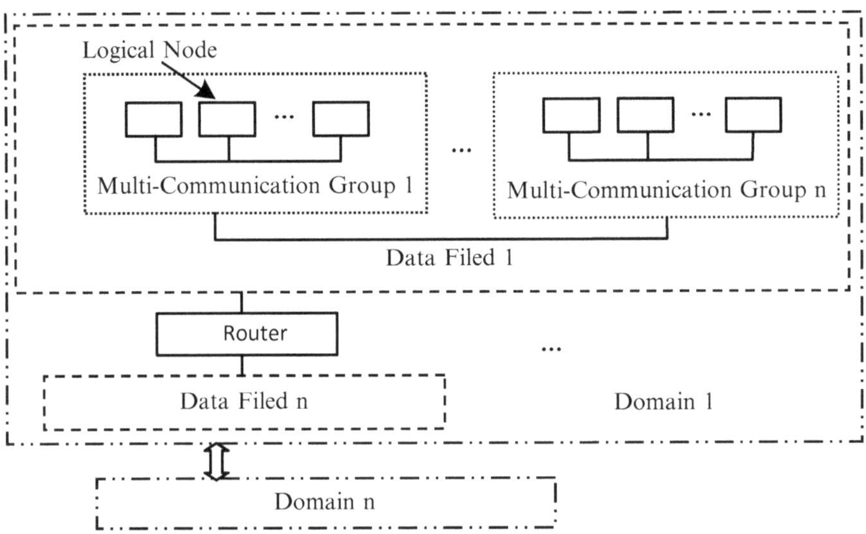


**Fig. 5.20** The construction frame of ADS (Reprinted from Ref. [7], Copyright 2006 with permission from IEICE)

## 5.4.2 *Design Principles*

The new diagnosis system of Maglev train should include some functions below.

### 5.4.2.1 Device Node

In Device Node, some fault information of single electric and electronic device should be produced by itself before sending status parameters to Device Node Layer. The correlative Device Nodes should be coordinated with each other. Corresponding fault message and integrated information should be sent to Device Node Layer except for simple status parameters of electric and electronic components. That is to say, single electric and electronic device in Maglev train should be independent and self-diagnosis. In addition, some corresponding electric and electronic devices should co-coordinate with each other, which is similar to Logical Node in autonomous decentralized system.

### 5.4.2.2 Device Node Layer

In Device Node Layer, corresponding electric and electronic devices should be divided into the same bus. Thus, some diagnosis function can be simplified. If some or a group of device nodes in the same Device Node Layer occur fault, the co-coordination mechanism of the whole Device Node Layer will start up. The fault message will be divided into several levels including single device node fault, type device node fault, and Device Node Layer fault. The corresponding fault message will be different based on different fault levels.

### 5.4.2.3 Fieldbus Layer

In Fieldbus Layer, different buses should be coordinated with each other, and the influence in them should be processed and embodied. That is to say, through the coordination of different buses, more complex fault information can be integrated and produced. For the whole Maglev train, the production of diagnosis message or fault message depends on not only single electric and electronic component but also several same devices or even the different type devices. Thus, the relations should be processed and embodied between different device node buses.

### 5.4.2.4 Diagnosis Layer

In Diagnosis Layer, all the status parameters, fault message, some fault information, and the whole diagnosis of Maglev train Vehicle-Section should be completed. That is to say, the Diagnosis Layer is the last layer of diagnosis system.

#### 5.4.2.5 Decision Layer

In Decision Layer, the decision function of the whole Maglev train should be completed. The fault information of single or a group of electric and electronic devices should be displayed, stored, and transmitted to Control Center.

#### 5.4.2.6 Logical Algorithms

The logical algorithms of diagnosis should be carried out in lower layer, which can improve the reliability and real time of system. In different layers of the diagnosis system, different logical algorithms will be circulated according to different diagnosis demands. In addition, these diagnosis algorithms should consider the whole demands including device maintenance, fault tolerance, and so on.

### *5.4.3 Design Plan*

Based on the design principles of diagnosis system, the frame of new diagnosis system of Maglev train is divided into four layers: Device Node Layer, Device Group Layer, Section Diagnosis Layer, and Train Diagnosis Layer. The configurations and functions of these layers are described below.

#### 5.4.3.1 Device Node Layer

This layer means single electric and electronic components. Its construction is shown in Fig. 5.21.

The design functions of Device Node include:

Acquire status parameters through the corresponding sensors in different locations of electric and electronic device and send status parameters. These status parameters can be sent to Device Fieldbus through CAN I/O Port in Device Node. Add diagnosis logical algorithm. Some fault messages can be gained by combining

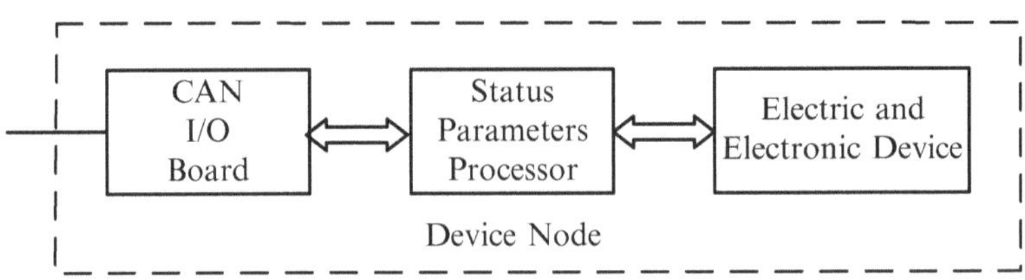


**Fig. 5.21** Device node frame (Reprinted from Ref. [7], Copyright 2006 with permission from IEICE)

different status parameters of electric and electronic device. Decide the fault of the whole device. If the result of logical judgment is true, a key fault device message can be produced and sent to Device Fieldbus through CAN I/O Port in Device Node.

#### 5.4.3.2 Device Group Layer

The division of Device Group mainly depends on the similarity and relativity of electric and electronic devices. All electric and electronic devices are divided into six Device Groups:

- Device Group of Magnet Levitation Control Units
- Device Group of Magnet Guidance Control Units
- Device Group of Magnet Brake Control Units
- Device Group of Energy Network Devices including Linear Generator Converters, Inverters, Energy Network Controllers, Energy Network Distributors, etc.
- Device Group of Location Measurement Units
- Device Group of other devices including Door Controller, Air-Conditioning Controllers, Fan Units, etc.

The frame of Device Group Layer is shown in Fig. 5.22.

In Device Group, the communication data includes status parameters from electric and electronic device in Device Nodes, fault device message from single Device Node, and Device Group fault information from Device Group Node. The diagnosis task of Device Group is completed in Device Group Node, which receives all fault device messages from different Device Nodes. Then the Device Group fault information can be produced through corresponding diagnosis algorithm. The Device Group Node only receives the messages about device fault, not including any status parameters from Device Nodes, which can reduce the load of Device Group Node.

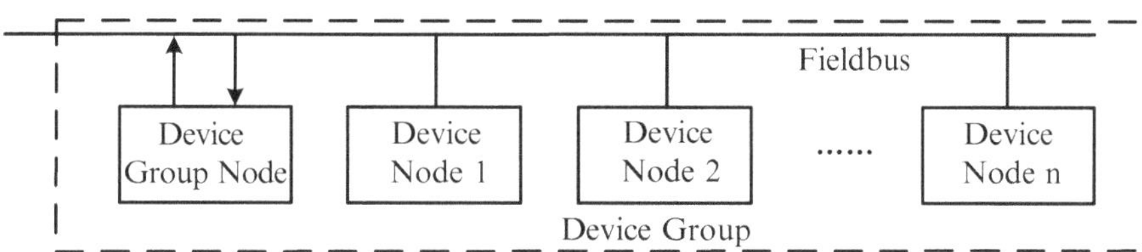


**Fig. 5.22** Device group frame (Reprinted from Ref. [7], Copyright 2006 with permission from IEICE)

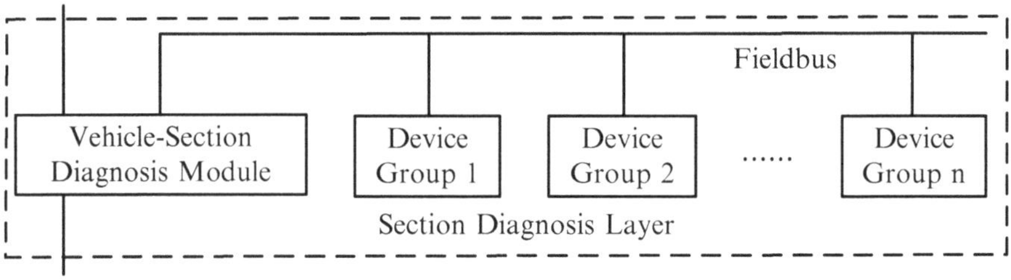


**Fig. 5.23** Section Diagnosis Layer frame (Reprinted from Ref. [7], Copyright 2006 with permission from IEICE)

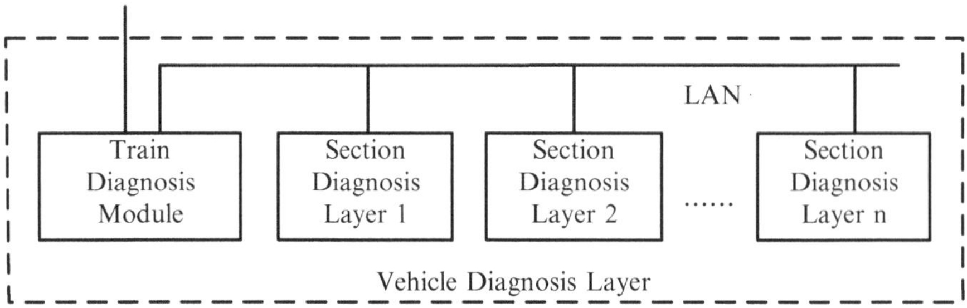


**Fig. 5.24** Train Diagnosis Layer frame (Reprinted from Ref. [7], Copyright 2006 with permission from IEICE)

#### 5.4.3.3 Section Diagnosis Layer

The Section Diagnosis Layer includes six Device Groups and Vehicle-Section Diagnosis Module in each Vehicle-Section of Maglev train. The frame of Section Diagnosis Layer is shown in Fig. 5.23.

In Section Diagnosis Layer, the communication data includes status parameters from electric and electronic device in Device Nodes, fault device message from single Device Node, Device Group fault information from Device Group Node, and Vehicle-Section fault information from Vehicle-Section Diagnosis Module. The diagnosis task of Vehicle-Section of Maglev train is completed in Vehicle-Section Diagnosis Module. It receives all fault messages from different Device Nodes and all Device Groups fault information from different Device Groups. The Vehicle-Section fault information can be produced through diagnosis algorithm in Vehicle-Section Diagnosis Module.

#### 5.4.3.4 Train Diagnosis Layer

The Train Diagnosis Layer includes several Section Diagnosis Layers and Vehicle Diagnosis Module in Maglev train. The frame of Train Diagnosis Layer is shown in Fig. 5.24.

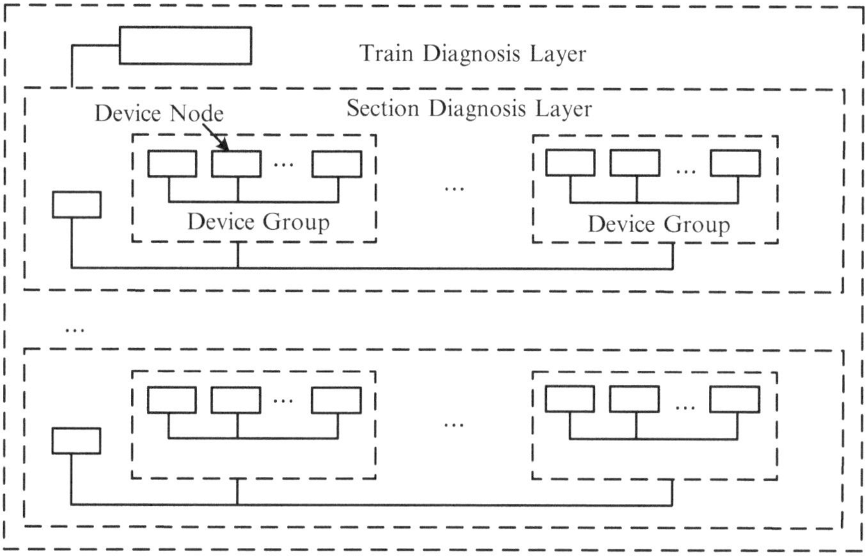


**Fig. 5.25** Diagnosis system frame (Reprinted from Ref. [7], Copyright 2006 with permission from IEICE)

In Train Diagnosis Layer, the communication data includes status parameters from electric and electronic components in Device Nodes, fault device message from single Device Node, Device Group fault information from Device Group Node, Vehicle-Section fault information from Vehicle-Section Diagnosis Module, and Train fault information from Train Diagnosis Module. The frame of whole diagnosis system of Maglev train is shown in Fig. 5.25.

### *5.4.4 Diagnosis Method*

In order to diagnose and judge the fault and status of electric and electronic devices in Maglev train correctly and quickly, the corresponding diagnosis methods should meet some demands below.

The production order of diagnosis message should be the status parameters of single device, single device fault, device group fault, device group combination fault, Vehicle-Section fault, and whole train fault. The diagnosis message of electric and electronic devices should be classified in terms of importance of message. During the course of fault and status diagnosis, the logical judgment algorithm should be carried out in proper means and location of diagnosis system. Diagnosis method should be decentralized into each level of diagnosis system and not into only one level of diagnosis system such as Vehicle Diagnosis Computer.

In old diagnosis system, the architecture levels can be divided into Device Node Layer, Fieldbus Layer, Section Diagnosis Layer, and Vehicle Diagnosis Layer. The level division is only based on actual devices and not based on diagnosis methods and system construction. The disadvantages of old construction have been discussed ahead.

In fact, all status parameters of all electric and electronic devices in Maglev train are transmitted to the Vehicle-Section Diagnosis Computer through two-level network.

Based on these status parameters of all electric and electronic devices in Maglev train, the diagnosis message of devices by themselves can be obtained with corresponding diagnosis methods. In addition, the diagnosis message of fieldbus can be obtained, too, which means that all status parameters from all kinds of devices without processing will be sent and all processing for status parameters will be centralized in Vehicle-Section Diagnosis Computer. It is obvious that the load of processing and transmitting message is largely added in Vehicle-Section Diagnosis Computer.

Of course, some ideas of ADS to improve the autonomous of each layer and device nodes can be adopted, and the processing task for status parameters can be separated into different layers. The old diagnosis method and new diagnosis method based on design plan above are compared and shown in Fig. 5.26a, b.

In Fig. 5.26a, b, it can be found that diagnosis functions of new diagnosis method are realized in lower layer than any of old method, which means that the fault messages and information will be produced more quickly and the load of Section Diagnosis Layer will be reduced largely. Specially, new method embodies the ideas

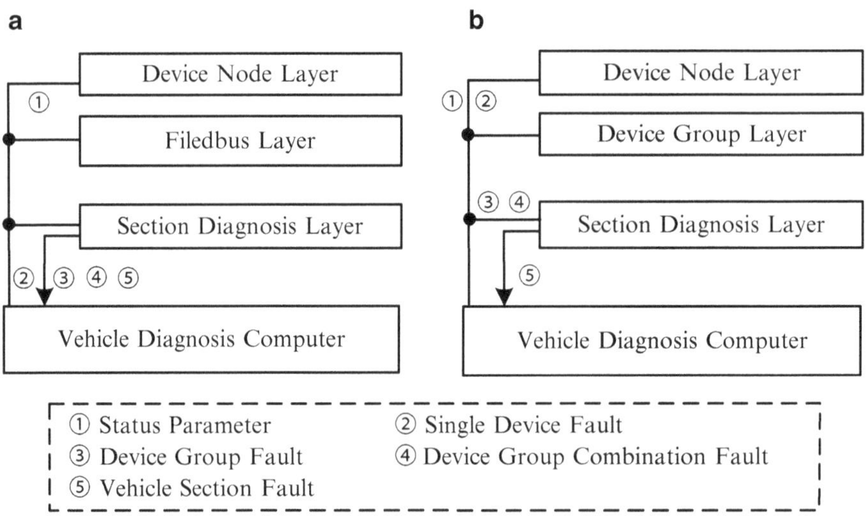


**Fig. 5.26** (**a**) Old diagnosis method, (**b**) new diagnosis method (Reprinted from Ref. [7], Copyright 2006 with permission from IEICE)

of ADS. If taken the Device Node as the example, self-control, self-coordination, and self-diagnosis functions can be processed.

In Device Group, the same and different device nodes can be co-coordinated and produce the local or whole diagnosis messages each other. From the ADS structure, the whole diagnosis system of Maglev train can be considered as an autonomous and decentralized system. Device Node, Device Group, and Section Diagnosis Layer are, respectively, similar to Logical Node, Data Field, and Domain in ADS.

The autonomous property of Device Node is ensured through Status Parameter Processor in Device Node. Device Node can initiatively send and receive messages from Device Group and other Device Nodes. The data consistency maintenance module can be added in Device Node. Thus, the function of fault tolerance will be realized in Device Node.

Meanwhile, considering the CAN Module's characteristic, the function of online expansion and online maintenance can be realized in Device Node, too. The autonomous of Device Group is ensured through Device Group Node in Device Group. Device Group can initiatively send and receive messages from Section Diagnosis Layer and other Device Groups. The harmonization and combination in Device Groups can be realized by adding corresponding algorithms in Device Group Node.

In Section Diagnosis Layer, the Vehicle-Section Diagnosis Computer can be considered as the router between the different Device Groups. The autonomous property of Section Diagnosis Layer is ensured through Vehicle-Section Diagnosis Computer. Of course, the Vehicle-Section Diagnosis Computer can be considered as the data storage device, too. All Device Nodes can initiatively send and receive messages based on their demands. The publish/subscribe architecture can be adopted in Section Diagnosis Layer.

It is not necessary that all status parameters and fault messages of electric and electronic devices will be displayed on the screen in driver room. But in old diagnosis system, all status parameters will be sent to Vehicle-Section Diagnosis Computer. Hence, these status parameters and fault messages must be gradually processed according to the layer construction in new diagnosis system. An example will be given to illustrate the advantages of new diagnosis method below.

It can found that the whole diagnosis algorithm flow of Energy Network should include four-level diagnosis algorithms in Fig. 5.27. High-level algorithms are based on low-level algorithms. These logical algorithms are all executed in Vehicle-Section Diagnosis Computer of old diagnosis system. In addition, the diagnosis of device groups not only depends on the status parameters of device groups by themselves but also is influenced by other device groups or device nodes. For example, the relation between Energy Network and Control Units is very close. The precondition of Control Units' good working state is that there are no faults for Energy Network. Inside device node, the condition is similar to device groups. That is to say, the relations in whole diagnosis system are very tight. In fact, the whole diagnosis algorithms cannot be simply described in Fig. 5.27. Hence, the diagnosis algorithms in old diagnosis system are complex and disordered.

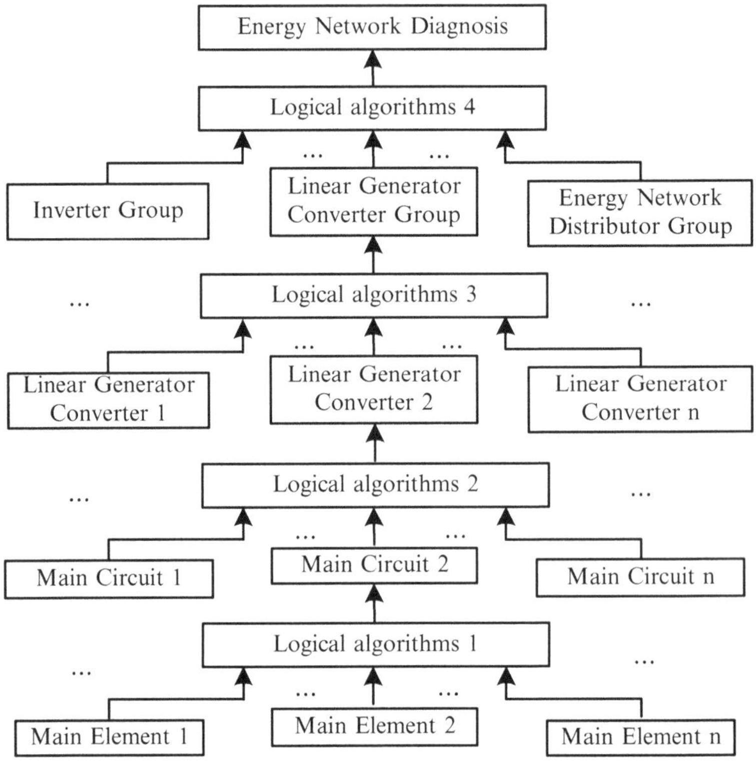


**Fig. 5.27** The diagnosis algorithm flow of energy network (Reprinted from Ref. [7], Copyright 2006 with permission from IEICE)

In new diagnosis system, the logical algorithms are separated into every layer based on the virtue of ADS' ideas, which means that the load of status processing parameters and transmitting messages in Vehicle-Section Diagnosis Computer is largely reduced and some important fault messages can be produced and transmitted more quickly without the processing in high layer.

The diagnosis algorithms in new diagnosis system are shown in Fig. 5.28. It can be found that the logical algorithms are properly assigned into each layer and are efficiently processed, which are only processed in Vehicle-Section Diagnosis Computer Layer for old diagnosis system.

If the transmission speeds of CAN and Ethernet are, respectively, 500 Kbps and 100 Mbps, in order to estimate the production time of fault message, the factors must be considered including computer CPU speed, CAN CPU speed, status parameter transmission time and delay, waiting time for status parameters, processing algorithms time, whole data quantity, and so on. The corresponding time can be estimated, for example, Energy Network Fault. The estimation results are shown in Table 5.3.

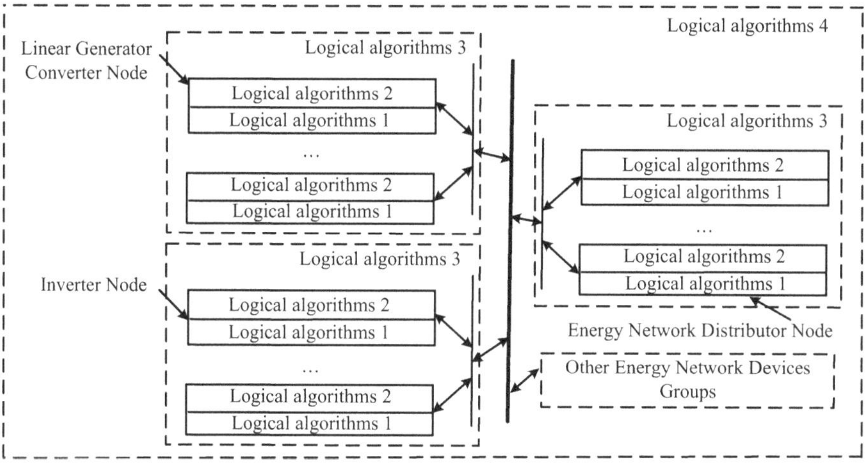


**Fig. 5.28** The diagnosis algorithms of Energy Network in new diagnosis system (Reprinted from Ref. [7], Copyright 2006 with permission from IEICE)

**Table 5.3** Time estimation of energy network fault

| Time (ms) | Old diagnosis system | New diagnosis system |
|---|---|---|
| Production time of status parameters | 0.01–0.03 | 0.01–0.03 |
| Transmission time | 0.002–0.005 | 0.002–0.005 |
| Processing time at node layer | 0 | 0.001–0.003 |
| Processing time at device group | 0 | 0.002–0.005 |
| Processing time (including delay time) in Vehicle-Section diagnosis computer | 0.04–0.05 | 0.01–0.02 |
| Total time | 0.052–0.085 | 0.024–0.063 |
| Load of Vehicle-Section diagnosis computer | 70–80 % | 30–40 % |

From the discussion above, it can found that the advantages of new diagnosis system plan and method are obvious. In fact, some ADS' ideas have been applied in the diagnosis system.

## 5.5 Control and Diagnosis Networks Based on Role Automation Decentralization

### *5.5.1 Relation of Control and Diagnosis Networks*

The control network and diagnosis network are independent from each other in Maglev train. There is only an interlace joint in Vehicle Control Units. In order to analyze and adjust the fault of electric and electronic devices, it is necessary to provide the supervisory signals of electric and electronic devices to diagnosis system. Similarly, the fault diagnosis of diagnosis network is also required to send and act some of control orders from control network.

As two important networks involving the safety of Maglev train, the coordination and cooperation between control network and diagnosis network are very important and necessary. However, the relationship and coupling between control network and diagnosis network are very poor. The message current about supervisory signals of electric and electronic devices from control network to diagnosis network is illustrated in Fig. 5.29. The message current about fault diagnosis of electric and electronic devices from diagnosis network to control network is illustrated in Fig. 5.30. From the discussion about the relationship of control and diagnosis networks in Maglev train, the cooperation between control network and diagnosis network can be completed through four different media transmissions, including CAN bus, hardware line, Ethernet, and TTY (current loop serial communication), which brings about the difficulty of maintenance, expansion, fault tolerance, dynamic change of control, and diagnosis functions.

In order to meet the demands including flexibility and cooperation of the whole control and diagnosis networks, real-time property of control orders and fault information transmission, online diagnosis of electric and electronic devices, correctness of control orders and fault message, and reliability of data communication, it is very necessary to design a whole network including all functions of control and diagnosis networks. The network should have online maintenance, online expansion, fault tolerance, dynamic change of control, and diagnosis functions.

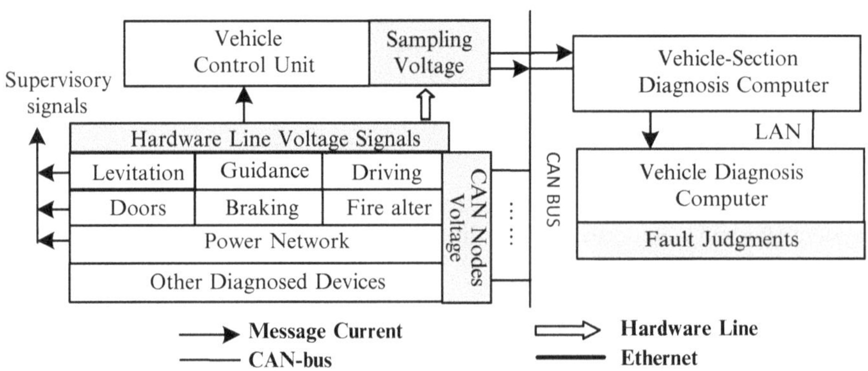


**Fig. 5.29** Message current of supervisory signals in electric and electronic devices from control network to diagnosis network (Reprinted from Ref. [8], Copyright 2006 with permission from IEICE)

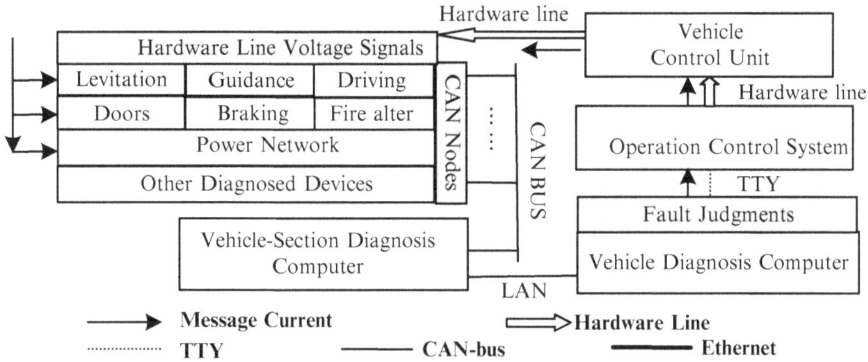


**Fig. 5.30** Message current of fault diagnosis in electric and electronic devices from diagnosis network to control network (Reprinted from Ref. [8], Copyright 2006 with permission from IEICE)

## 5.5.2 The Concept of RoADS

The key technology of ADS includes architecture technology based on data field and content code communication technology based on publish/subscribe model. With the further study about ADS and its corresponding theory, traditional automation systems gradually have more demands for their applications and development. The requirements for space time dynamic evolution representation model of automation system potential functions are coming out. The space time dynamic evolution representation model of system potential functions corresponds to a kind of special automation system technology. The technology is called role automation decentralized system (RoADS) [26, 27]. The corresponding concepts are defined as follows:

Role: it means special function which is external expression of object in the special time.

Agent: it means a node or subsystem. Its elements include controllers, software, sensors, and actors.

Automation system: it means the group of several agents. They can cooperate with each other to realize the goal of system.

Role of automation system: it means special function which is external expression of an agent in the special time. Several roles of several agents consist of the role set of the automation system.

Role autonomy: it means the ability of different functions' independent expression for an agent according to the change of internal status or external conditions in different time and places.

Role automation decentralized system: it means the automation system consisted of several agents with role autonomy. The agent has some properties as the following:

Function equity: every agent has the potential equal functions.

Role autonomy: agent can independently express function according to status and conditions.

Role coordination: for nonworking status of every agent, other agents can exert its role according to coordination of different goals.

Role Field and data field dual-driven architecture is presented as follows:

The separation between the transmission strategy (anonymity and asynchronism) of content code (CC) data and data processing mechanism (subscribe and filter) in ADS realizes the online dynamic characteristic of system. In role automation decentralized system, the role expression actually means the dynamic recombination course of a function set. The separation between role description strategy (functions expression) and role control mechanism (functions recombination) is expected, which will realize the dynamic evolution of roles with the change of time space. In the most automation systems, the role description strategy (functions expression) and role control mechanism are synergetic, which results in the determinate expression functions of the whole system in the design phase. The role field and data field dual-driven architecture (RDDA) is named in [27]. In the architecture of RDDA, the main elements include role field, data field, role autonomous processor, data arbitrator, role/function/object dispatcher, and so on. All agents in the system have the same structure. The architecture of RDDA is illustrated in Fig. 5.31.

#### 5.5.2.1 Role Field

The autonomous agent embedded in a system frame can play several roles at same time. In addition, the role sets of the agent can dynamically change with time. The system frame is called role space. The role is available to the agent. Through the roles, agent can complete single or same goals. The role field is sharable information

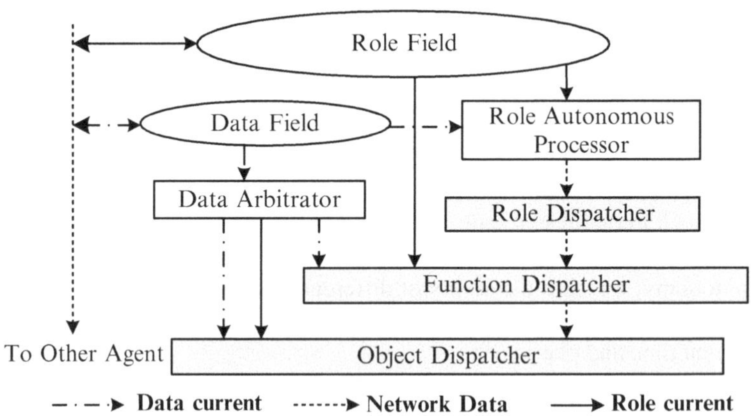


**Fig. 5.31** Architecture of RDDA (Reprinted from Ref. [8], Copyright 2006 with permission from IEICE)

space. Every agent can send or receive role description information to the role field. The description information of the role includes role component files, function component files, and object component files.

#### 5.5.2.2 Data Field

The data field is sharable data space. Every agent can send or receive data to the data field. Data includes the properties and examples of objects, internal and external status, and so on. The internal status includes resource status, health status, disposal ability, and so on. The external status includes local event, remote event, communication event, change of control object, and so on. Data current is formed through the exchange of the agents.

#### 5.5.2.3 Role Autonomous Processor

The role autonomous processor is the core elements of RoADS' realization. It can get the role component files from role field, get data from data field, arbitrate the current running roles of the agent, drive the corresponding role components, and activate the acting role. From the information viewpoints, there are data current and role current in RoADS. At the level direction (between the agents), the consistency of data and role should be achieved. At the upright direction (among the agents), the association of data and role should be achieved.

#### 5.5.2.4 Data Arbitrator

Each agent can run in decentralized conditions. There is a great deal of data for the transmission and updating timely. Considering the uncertainty of data during the course of transmission, data arbitrator can ensure the consistency of data in decentralized conditions.

#### 5.5.2.5 Role/Function/Object Dispatcher

Role dispatcher is driven by role autonomous processor to realize the circumstance configuration of role activation. It can have alternation with the control interface and realize the automatic or constrained switch. Function dispatcher is driven by role dispatcher, function component files, data and control interfaces to realize the function configuration of the role, and activation of corresponding function component. Object dispatcher is driven by role dispatcher, object component files, data and control interfaces to realize the object configuration of the function, and activation of corresponding objects.

In virtue of the RoADS idea, the control and diagnosis networks in Maglev train will be newly designed and realized below in this section.

## *5.5.3 Design Principles*

The new control and diagnosis network in Maglev train should include some basic design principles below.

### 5.5.3.1 One Network

It means that the new network should replace old control and diagnosis networks in Maglev train. The new network should include and complete all functions of old control and diagnosis networks. The hardware lines network should be completely given up, and the new network should be intelligent network with better communication characteristic.

### 5.5.3.2 Distinct Priority

Since the diagnosis and control messages are transmitted in one network, it is very necessary that there should be distinctly priority different between diagnosis and control messages, for the sake of ensuring the timely and correct transmission of control signals.

### 5.5.3.3 Information Share

Different from the separation of control and diagnosis networks, all information of two networks can be shared. Thus, for the judgment of diagnosis and fault, control orders and supervisory signals should provide the significance support and reference. Before sending the control orders and judgment of supervisory signals, diagnosis information should strongly provide help and foundation.

### 5.5.3.4 Function Change

Considering the redundancy of diagnosis and control systems in Maglev train, the fault of any electric and electronic devices (including Magnet Levitation Control Unit, Magnet Guidance Control Unit, Magnet Brake Control Unit) cannot result in the traveling halt. When one or some electric and electronic devices come forth the fault, other same type nodes should own the functions of fault nodes. For the diagnosis information, similar functions should exist for diagnosed devices.

#### 5.5.3.5 Role Change

In all electric and electronic devices of Maglev train, some devices act double roles. On one hand, the devices receive the request frame from diagnosis system and send status information to diagnosis network. On the other hand, they receive the control orders from control system and send supervisory signals to control network. These devices mainly include Magnet Levitation Control Unit, Magnet Guidance Control Unit, Magnet Brake Control Unit, and so on. Their double roles should provide the functions of role change in the new network.

#### 5.5.3.6 Hierarchy Diagnosis

Being different from the control message, the diagnosis message should be divided into different levels for the sake of diagnosis convenience.

In Device Node, some fault information of single electric and electronic device should be produced by itself before sending status parameters to Device Node Layer. The correlative Device Nodes should be coordinated with each other. Corresponding fault message and integrated information should be sent to Device Node Layer except for simple status parameters of electric and electronic device. That is to say, single electric and electronic device in Maglev train should be independent and self-diagnosis. In addition, corresponding electric and electronic devices should co-coordinate with each other, which is similar to Logical Node in autonomous decentralized system.

In Device Node Layer, corresponding electric and electronic devices should be divided into the same bus. Thus, some diagnosis function can be simplified. If some or a group of device nodes in the same Device Node Layer cause fault, the co-coordination mechanism of the whole Device Node Layer will start up. The fault message will be divided into several levels including single device node fault, type device node fault, and Device Node Layer fault. The corresponding fault message will be different based on different fault levels.

In Fieldbus Layer, different buses should be coordinated with each other, and influence in them should be processed and embodied. That is to say, through the coordination of different buses, more complex fault information can be integrated and produced. For the whole Maglev train, the production of diagnosis message or fault message depends on not only single electric and electronic device but also several same devices or even the different type devices. Thus, the relations should be processed and embodied between different device node buses.

In Diagnosis Layer, all the status parameters, fault message, some fault information, and the whole diagnosis of Maglev train Vehicle-Section should be completed. That is to say, the Diagnosis Layer is the last layer of diagnosis system.

In Decision Layer, the decision function of the whole Maglev train should be completed. The fault information of single or a group of electric and electronic devices should be displayed, stored, and transmitted to Control Center.

The logical algorithms of diagnosis should be carried out in the lower layer, which can improve the reliability and real time of system. In different layers of the diagnosis system, different logical algorithms will be circulated according to the different diagnosis demands. In addition, these diagnosis algorithms should consider the whole demands including the device maintenance, fault tolerance, and so on.

## 5.5.4 *Design Plan*

Based on the introduction and discussion of role automation decentralized system (RoADS), and the design principles of control and diagnosis network in Maglev train, the frame of new network is divided into several parts including agents, roles, data field, role field, role field and data field dual-driven architecture, and system frame. Each part will be described and discussed in detail below.

### 5.5.4.1 Agent Structure

In this chapter, the agent means the every node in control and diagnosis network including Magnet Levitation Control Unit, Magnet Guidance Control Unit, Magnet Brake Control Unit, Linear Generator Converter, Energy Network Distributor, Energy Network Controller, Vehicle Control Unit, Door Controller, Location Measurement Unit, Air-Conditioning Controller, Inverter, Fan Unit, and so on. For the sake of convenience, CAN bus is adopted in the new network.

That is to say, all control orders, supervisory signals, and diagnosis information will be transmitted through CAN bus. An agent is generally consisted of system engines such as interface configuration and edit, properties and logical control, data packing and analysis, network communication middleware, and configuration files transmission. The structure of agent is illustrated in Fig. 5.32.

The configuration files are produced through corresponding objects, devices' properties, and action logical conditions. The validity of control order and diagnosis message is checked by property and logical control engines to realize the sequential operation in property and logic. Data packing and analysis engine can pack the data produced by property and logical control or analyze the data from the CAN bus network middleware. The autonomous decentralized protocol (ADP) is adopted in CAN bus communication middleware. The files transmission engine can automatically receive and send configuration files between device node and network controller.

Through the device node, the status parameters and supervisory signals through corresponding sensors in different locations of electric and electronic device can be obtained. The control orders from network controller can be checked and received in the corresponding agent. The status parameters and supervisory signals can be sent to Device Fieldbus through CAN I/O Port in Device Node. Some fault messages can be gained by combining different status' parameters of electric and electronic

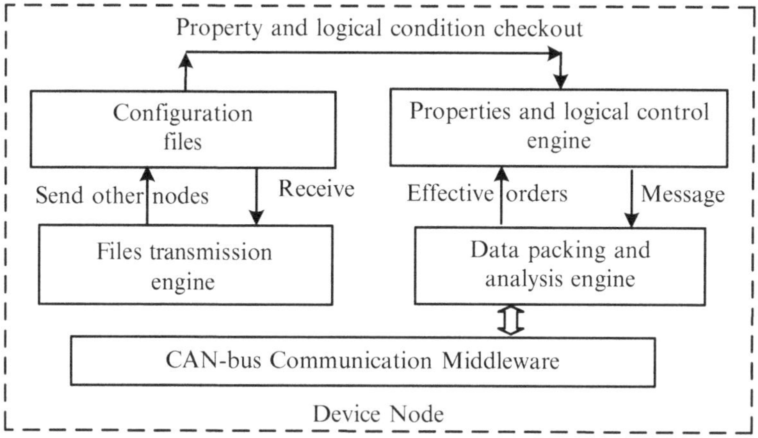


**Fig. 5.32** Structure of agent (Reprinted from Ref. [8], Copyright 2006 with permission from IEICE)

device. If the result of logical judgment is true, a key fault device message can be produced and sent to Device Fieldbus through CAN I/O Port in Device Node.

In the control and diagnosis network of Maglev train, all devices nodes have same agent structure.

#### 5.5.4.2 Role and Role Field

In the control and diagnosis network in Maglev train, the role means special function which is external expression of electric and electronic device in the special time. For example, during the traveling course of Maglev train, the functions of every electric and electronic device are different.

Depending on the similarity and relativity of roles, all electric and electronic devices are divided into 2 sets. They are, respectively, Diagnosis Role Set (Magnet Levitation Control Units, Magnet Guidance Control Units, Magnet Brake Control Units, Energy Network Devices, Location Measurement Units, other devices including Door Controller, Air-Conditioning Controllers, and Fan Units) and Control Role Set (Magnet Levitation Control Units, Magnet Guidance Control Units, Magnet Brake Control Units, Energy Network Devices, Location Measurement Units, other devices including Door Controller, Air-Conditioning Controllers, and Fan Units).

The electric and electronic device can own several roles at the same time, such as Magnet Levitation Control Units. Moreover, the role sets of the electric and electronic device can dynamically change with time. The control and diagnosis network frame can form the role space. Through the roles, electric and electronic device can complete single or same goals. Every electric and electronic device can send or receive role description information to the role field. The description information of the role includes role component files, function component files, and device component files.

#### 5.5.4.3 Data Field

In the control and diagnosis network in Maglev train, data field means the sharable data space. In fact, it is the memory of network controller. Every electric and electronic device can send or receive data to the data field. Data includes the properties and examples of electric and electronic device, internal and external status, and so on.

The internal status includes node's resource status, node's health status, node's disposal ability, and so on. The external status includes local network event, remote network event, network communication event, network change of control device, and so on.

#### 5.5.4.4 RDDA

In the control and diagnosis network of Maglev train, all electric and electronic devices have the same agent structure. The architecture of RDDA is illustrated in Fig. 5.33.

In Fig. 5.33, Role dispatcher is driven by role autonomous processor to realize the circumstance configuration of role activation. It can have alternation with the control interface and realize the automatic or constrained switch.

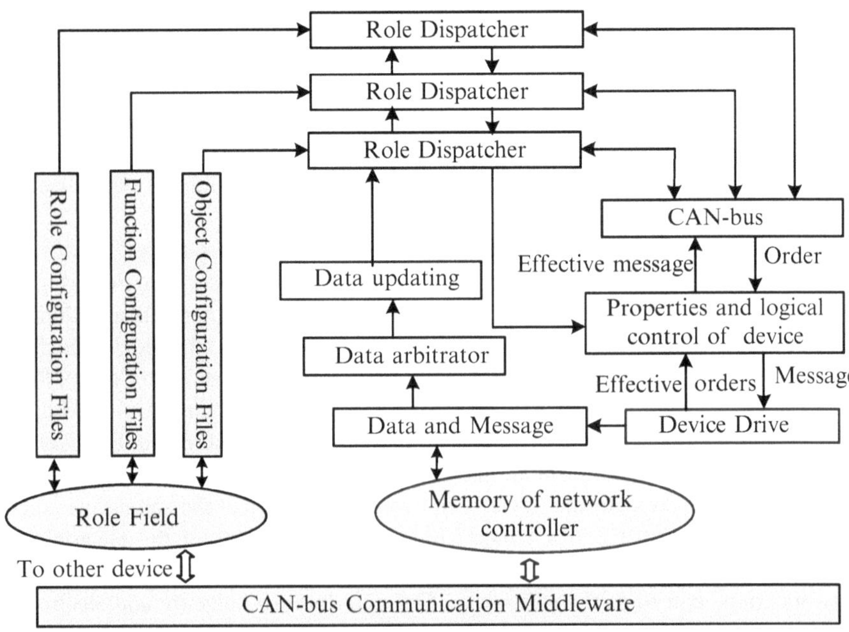


**Fig. 5.33** Architecture of RDDA in control and diagnosis network (Reprinted from Ref. [8], Copyright 2006 with permission from IEICE)

The function dispatcher is driven by role dispatcher, function component files, data, and control interfaces to realize the function configuration of the role and activation of corresponding function component.

Object dispatcher is driven by role dispatcher, object component files, data, and control interfaces to realize the object function configuration and activation of corresponding objects.

#### 5.5.4.5 System Frame

The frame of control and diagnosis network in Maglev train is shown in Fig. 5.34.

Considering computer CPU speed, CAN CPU speed, status parameter transmission time and delay, waiting time for status parameters, processing algorithms time, the whole data quantity, and so on, the corresponding time can be estimated. Suppose the transmission speed of CAN be 500 Kbps.

In addition, the performance comparison between new and old networks is shown in Table 5.4.

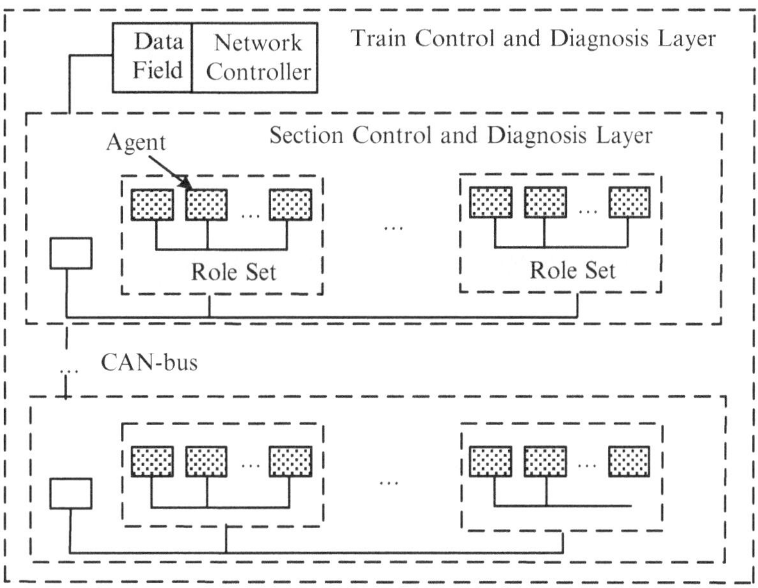


**Fig. 5.34** Frame of control and diagnosis network (Reprinted from Ref. [8], Copyright 2006 with permission from IEICE)

**Table 5.4** The performance comparison of new and old networks

| Performance | Old networks diagnosis/control | New network |
|---|---|---|
| Production time of status parameters (ms) | 0.01– 0.03/0 | 0.01–0.03 |
| Total time (ms) | 0.052–0.085/0 | 0.024–0.065 |
| Hierarchy diagnosis | No/no | Yes |
| Distinct priority | No/no | Yes |
| Information share | No/no | Yes |
| Function change | No/no | Yes |
| Role change | No/no | Yes |

## 5.6 Onboard Communication Platform in Maglev Train Based on RTLinux

### *5.6.1 RTLinux Introduction*

RTLinux is one kind of real-time operating systems based on Linux. On the one hand, it has the hard real-time mechanism, and on the other hand, it can own all the advantages of Linux itself. For the essence of RTLinux, a micro-kernel real-time process, also known as the RTLinux real-time subsystem, is designed based on original Linux [28]. Namely, a virtual machine layer (RTLinux kernel) is inserted between the Linux kernel and the hardware interrupt. Meanwhile, Linux system is applied as a low priority task in RTLinux operates.

The frame of RTLinux is shown in Fig. 5.35. It is known that the task switching time and task interruption time of several real-time operating systems (RTLinux3.2, VxWorks, QNX6.1, WinCE) are about 10 μs, which satisfy the hard real-time demand of onboard communication platform in Maglev train. Compared with other real-time operating systems, RTLinux has some advantages, such as perfect "Anjuta" and "Kdevelop" integrated development environments, high performance "gcc" compiler and "Autotools" compilation tools, and free and open original code. Hence, based on the real-time property, system portability, and development cost, RTLinux is very suitable to onboard communication platform in Maglev train.

### *5.6.2 Design Plan*

Due to the special requirements of high-speed operation, electromagnetic interference, and frequent vibration for Maglev train, CPCI (Compact Peripheral Component Interconnect) industrial computer is adopted as the onboard communication platform. The network of four CAN buses and Ethernet is still adopted. Red Hat Linux 2.4.20 and RTLinux 3.2 are adopted as the software development platform.

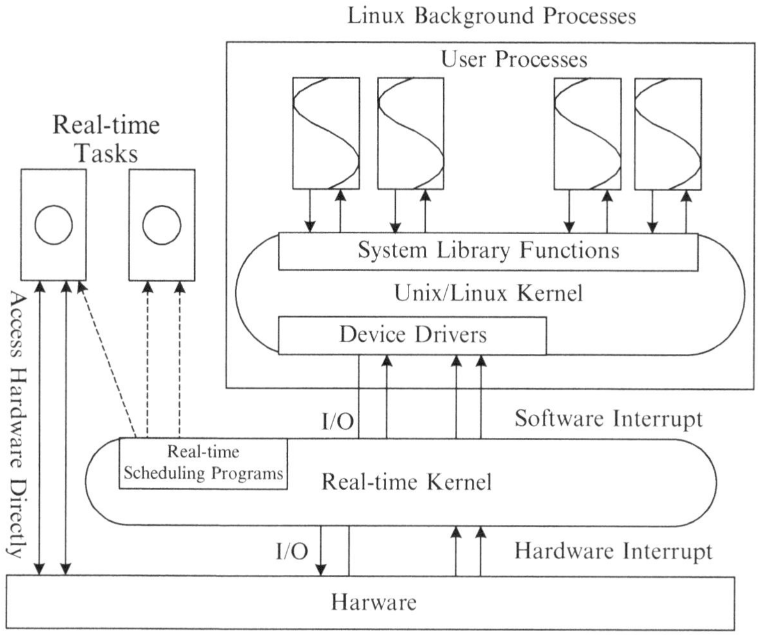


**Fig. 5.35** Frame of RTLinux

#### 5.6.2.1 Communication Platform Structure

The onboard communication platform software can be divided into three modules based on the function demands of diagnosis system in Maglev train and basic structure of RTLinux:

① Drive module for CPCI-CAN card driver.
② Real-time processing module that runs in kernel space communication platform will mainly realize real-time thread, create RT-FIFO, and share memory.
③ Non-real-time processing module that runs in the user space will mainly realize the communication platform initialization tasks and send communication packets.

The system architecture is shown in Fig. 5.36.

The real-time part mainly completes the status data acquisition of underlying components in four CAN buses and monitors the status of each node. In order to be able to run with minimal time, the real-time part is small and simple.

The non-real-time mainly completes the self-check of communication platform, resource allocation management, initialization of the CAN devices, protocol conversion between CAN and Ethernet, and communication with train diagnosis and operation computers. The real-time process is created through a kernel module initialization of RTLinux,which will realize the data acquisition of CAN and

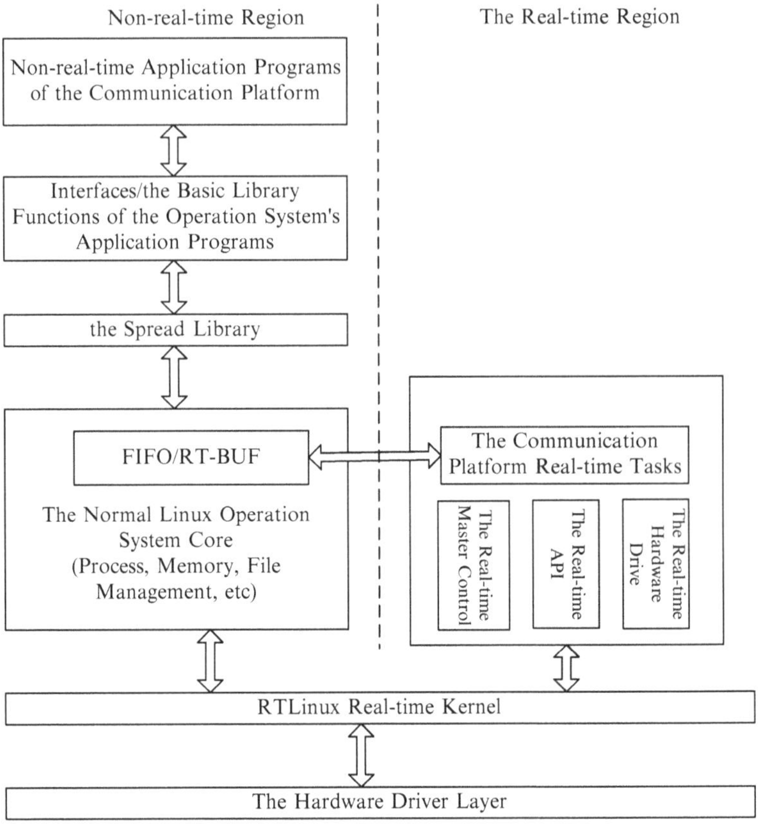


**Fig. 5.36** System architecture of communication platform (Reprinted from Ref. [15], Copyright 2006 with permission from IEICE)

data transmission to the non-real-time processing module by sharing memory and receiving the control commands through RT-FIFO from user space. The detailed process is listed below.

Firstly, the real-time module runs "init_module()", creates two RT-FIFOs, and sets the corresponding interrupt handler real-time tasks of writing RT-FIFO1 operation.

Secondly, the RT-FIFO1 control command is written to run the initial real-time task from the main program running in user mode.

Lastly, main program will monitor RT-FIFO2, wait for the data acquisition of real-time task 2, and complete the processing of signals. In real-time task 1, the thread property is set and real-time thread 1 is created. If the running period is set at the start point of real-time thread 2, real-time thread 2 will run according to the period.

#### 5.6.2.2 Real-Time Processing Module

The standard frame of CAN bus is adopted for the data communication, namely, independent CAN controller SJA1000 runs under "BasicCAN" mode. The real-time control module codes of onboard communication platform in Maglev train mainly include three sections: communication platform module initialization codes, communication platform real-time task codes, and communication platform download codes.

Module initialization of communication platform software: different from the general application that must have a main function, there is no main function in kernel module programming of RTLinux.

The start point "init_module()" of kernel module is called when it is loaded into the kernel and completes the parameter initialization.

The main function of "init_module()" in real-time processing module of onboard communication platform software is listed below:

Hardware initialization of CPCI-CAN card: channel selection and control register programming of CPCI-CAN card

Real-time thread creation: real-time task thread creation through function "pthread_create()" and thread priority setting

RT-FIFO creation: control command bytes transmission between real-time processes and non-real-time processes

Memory share: large quantities of data transmission between real-time processes and non-real-time processes

Real-time tasks of communication platform: the real-time tasks mean the development of some communication platform threads including communication scanning thread, communication reading thread, communication forwarding thread, and so on.

For example, the scanning thread is responsible for checking the received information. If a node doesn't receive information for long time, the thread will send the single-node inquiry message, and the time when "read thread" is putting into share memory is acted as the start time point, which is shown in Fig. 5.37.

Communication platform module download: the created RT-FIFO, sharing memory, and real-time thread are initialized through communication platform. The functions that release their space are called to realize the module download, and these functions can be called with "cleanup_module()".

#### 5.6.2.3 Non-Real-Time Processing Module

The main thread in non-real-time processing modules finishes all initialization tasks, and broadcasting and control messages are periodically sent to four CAN ports. The initialized process includes the driver installation of CPCI-CAN card, the buffer

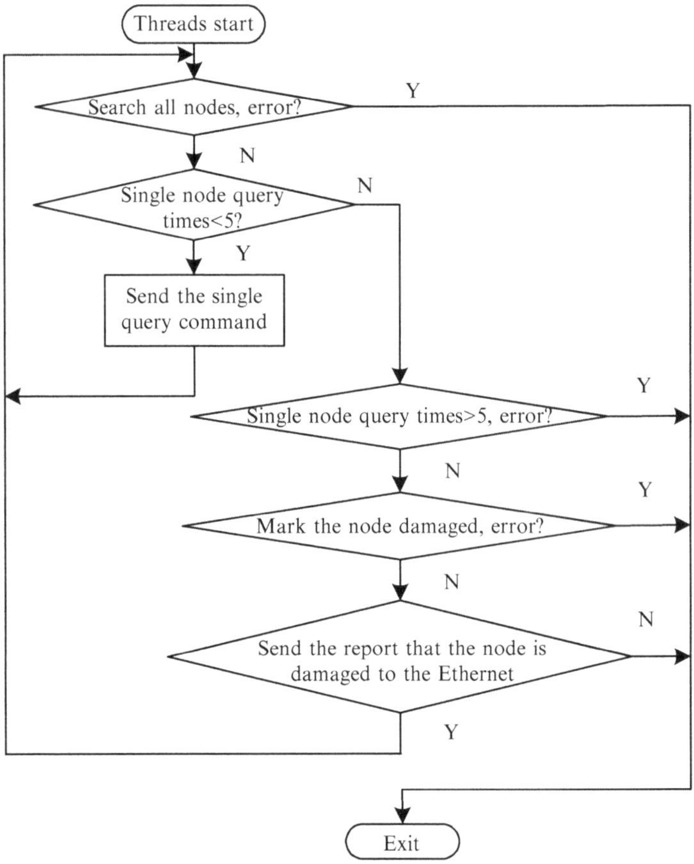


**Fig. 5.37** Scan thread flow chart (Reprinted from Ref. [15], Copyright 2006 with permission from IEICE)

creation for receiving CAN message, and the opening of four CAN ports. The initialization of CAN control chips includes checking hardware connection; entering reset state; setting clock frequency demultiplication register; outputting control register, communication baud rate, code acceptance register, and code shielding register; exiting the reset mode and setting work mode; and setting interrupt to enable register.

For Ethernet, the non-real-time processing module is responsible for receiving the connection of TCP client and sending control message to four CAN ports when receiving the control message.

### 5.6.3 Driver Design of CPCI-CAN Card

Considering that the normal work of CPCI-CAN card is the basic condition that ensures the smooth running of Maglev train onboard communication platform, the driver of CPCI-CAN card based on RTLinux must be developed. In RTLinux, the non-real-time driver is same as that in Linux. However, the driver in Linux cannot be ensured to run in RTLinux. For example, the standard Linux device drivers often disable interrupts, while the task in RTLinux non-real-time kernel cannot really disable interrupts, which may result in the driver's operation error. In addition, the application of non-real-time domain in RTLinux can be stolen by the real-time tasks. In order to make full use of real-time characteristics of RTLinux hardware, the appropriate driver must be completed in the form of real-time tasks. The initialization flow chart of CPCI-CAN card is shown in Fig. 5.38. Its driver includes registration module and interface module. The realization process is listed below.

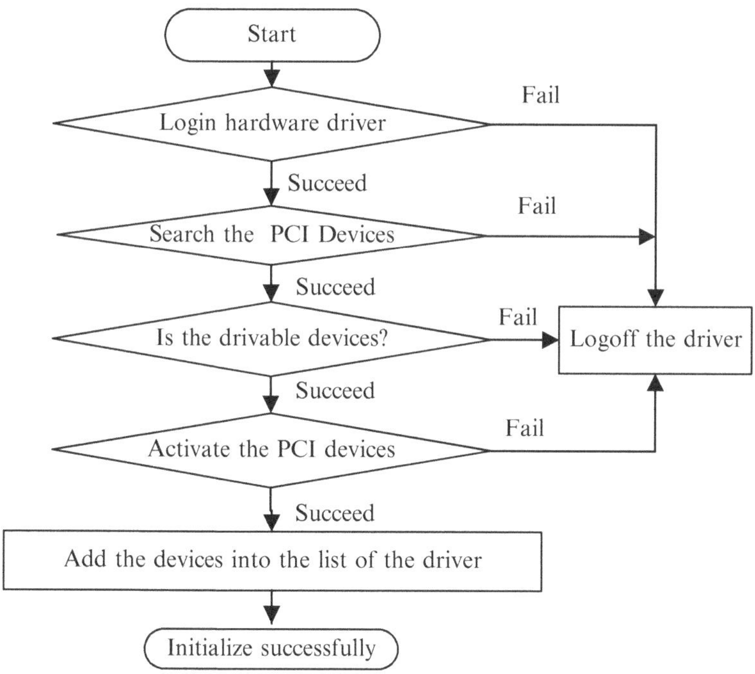


**Fig. 5.38** Flow chart of PCI-CAN initialization (Reprinted from Ref. [15], Copyright 2006 with permission from IEICE)

#### 5.6.3.1 Registration Module

Firstly, the device will be found according to the PCI specification, and CPCI-CAN card will be checked in "init_module()" module. If the module does not exist, the load of device driver will exit; otherwise the resource and interrupt number are assigned to the found device. Meanwhile, real-time FIFO is created, the real-time processing function is registered in real-time FIFO. "register_chrdev()" function is used to register communication card driver in the system and will apply the interrupt for CPCI-CAN card. If these resources application is not successful, the system resources will be released and exit from the loading of device driver. In "cleanup_module" module, "cleanup_module" function is firstly used to release the applied interrupt in "init_module" module.

Secondly, "kfree()" function is used to release the applied memory space in "init_module" module.

Thirdly, "unregister_chrdev()" function is used to release the registered driver in "init_module" module. The simple codes of register hardware driver and register character device driver are listed below.

```
/* register hardware driver */
rc = pci_register_driver(&p7841_driver);
if(rc < =0)
{printk(KERN_ALERT"Can't register p7841
as a pci device!\n");
pci_unregister_driver(&p7841_driver);
return-ENODEV;}
/* register character device driver */
rc = register_chrdev(MAJOR_NUM,"p7841",
&p7841_fops);
if(rc < 0){
printk(KERN_ALERT"Can't register p7841
as a chr device!\n");
return rc;}
```

#### 5.6.3.2 Driver Interface Module

The driver interface module provides user with the interface of operation object devices. The main functions "P7841_open()" and "p7841_close()" are called to complete the opening and releasing devices. The device opening is completed through calling "open()" function in "file_operations" structure and the global registers are configured. Real-time FIFO is used to finish the communication of real-time and non-real-time parts. The configuration of global registers mainly includes the receiving configuration of devices, writing transmit buffer, and reading receive buffer.

The device releasing is completed through calling “release()” function in “file_operations” structure. In addition to read and write operations, it is necessary to control equipment, which can be finished through “ioctl()” function in device driver. After the driver is programmed, the drive module is loaded into the kernel by calling “insmod()” command. In addition, the kernel of Linux can also directly be loaded into the kernel through the recompile.

## 5.6.4 *Real-Time Performance Analysis*

### 5.6.4.1 Real-Time Performance Test

The real-time performance of RTLinux can be tested with the dispatch time that is the cost time during the course of real-time task dispatch in system. The test program includes two modules of real-time task and non-real-time task. The real-time task is mainly responsible for the different computation between actual dispatch point and expected dispatch point and writing the different value into FIFO device [29]. The main function is an infinite loop, which is responsible for inputting maximum and minimum testing values of 1,000 times into FIFO device. The non-real-time task will read these values from FIFO device.

```
/* real-time task */
int init_module(void)//initialize module
{fifo_status rt_create (0.4000);
//create real-time FIFO for communication
thread_status = pthread_create(&taskl,&attr,
thread_ code,(void *)1);
// create a real-time task thread}
void cleanup_module(void)//download module
{pthread_delete_np(taskl);
        close(fifo);
        rtf_destroy(0);}
/* non-real-time task */
for(;;){
read(F1F0),max,min);
printf("%d, %d\n",max,min);}
```

The two parts are respectively compiled. The real-time processing modules are inserted into the kernel and the program of non-real-time task runs. The period of each loop is 1 ms and the single trigger means is adopted. CPU is 1.7 GHz, 10 group test data, the maximum dispatch time is 12.320 μs, and the minimum dispatch time is 12.192 μs. Based on the testing results, the dispatch time is about 12 μs, which can satisfy the real-time performance demand of onboard communication platform in Maglev train.

#### 5.6.4.2 Real-Time Performance Analysis

The real-time of onboard communication platform in Maglev train mainly depends on the hardware performance in CAN bus, communication protocols, network load, and the channel utilization rate. The main real-time influence factors include the delay of sending packets, minimum interrupt time of CAN bus, response speed of CPCI-CAN card, and the processing speed of host computer when the underlying electrical and electronic components in onboard diagnosis system communicate with diagnosis computer.

In order to analyze the real-time of communication platform, the general transmission delay model is adopted in Fig. 5.19. Supposing that the number of underlying electrical and electronic components is 132 and all CAN nodes are evenly distributed to four CAN buses, the packets produced by node components are serially sent to the vehicle diagnosis computer in onboard diagnosis communication platform. The actual completion time of each node sending packet is equal to the sum of clock time, interrupt time, and transmission time of vehicle diagnosis computer, which is shown below.

$$T = T_{\text{delay}} + T_{\text{irrupt}} + T_{\text{clk}} \tag{5.8}$$

Based on Fig. 4.1, $T_{\text{delay}} = T_{\text{d}} - T_{\text{s}} = T_{\text{queue}} + T_{\text{block}} + T_{\text{frame}} + T_{\text{trans}} + T_{\text{rec}}$. There are generally old packets in CAN bus due to the real-time demand in CAN and $T_{\text{queue}}$ can be ignored. In addition, considering the fact that the transmission distance of CAN bus in Maglev train is short, $T_{\text{trans}}$ and $T_{\text{rec}}$ can be also ignored. Hence, $T_{\text{delay}} \approx T_{\text{block}} + T_{\text{frame}}$. $T_{\text{block}}$ is decided by the MAC access mechanism. $T_{\text{frame}}$ depends on the frame length and bit time of packets. If the baud rate is 500 kb/s and there are 8 bytes in each node packet, the general transmission delay can be computed as follows under the maximum load operation:

$$T_{\text{delay}} = (44 + 8 \times 8)/500 = 0.216\text{ms} \tag{5.9}$$

$$T_{\text{irrupt}} = 3/500 = 0.006\text{ms} \tag{5.10}$$

$T_{\text{clk}}$ is the clock time of host computer. If the program execution time of host computer is ignored, the refresh time of each CAN bus is

$$T_{\text{all}} = (0.216 + 0.006 + 10) \times 30 = 306.66\text{ms} \tag{5.11}$$

Based on the design of onboard diagnosis communication platform in Maglev train, the response time is about 500 ms. Through the actual test and data analysis, the average refresh time of each CAN is about 290 ms.

## 5.7 Conclusion

In the first section, the simulation model of onboard diagnosis network of Maglev train with OPENT is built to analyze the network's performance. Based on the simulation results, BER is the most important factor for the onboard diagnosis network performance, and the influence of response error for the network performance is small. Considering that the actual onboard diagnosis network of Maglev train is a complex system, the actual CAN node models are simplified for the simulation. There are some limitations for the performance study of actual device nodes, and the approaching work will focus on the node model improvement and parameter setting, which will provide better support for the performance of actual onboard diagnosis network of Maglev train.

In the second section, for the control network of Maglev train, the hardwire network (each signal is represented by two hardwires) is adopted. The network has some disadvantages such as poor interconnection and expansibility, mass redundancy, difficult wire layout, much weigh addition for Maglev train, and so on. Through the analysis and comparison of mainstream train communication networks (MVB/WTB, CAN, Lonworks, ARCnet) on high-speed train, CANOpen technology is adopted to integrate the control and diagnosis networks of Maglev train. The equipment and its application of new integrated network based on CANOpen are discussed and designed. The communication properties of the new integrated network based on CANOpen for Maglev train can be studied.

In the third section, the traditional supervision and control system are adopted in the diagnosis system. Due to many disadvantages of the diagnosis system, some ADS' ideas are adopted in new diagnosis system in this paper. The design plan and diagnosis method are proposed, discussed, and analyzed in detail. Through comparison between old diagnosis system and new one, it is concluded that new diagnosis system can better meet the demands of Maglev train diagnosis and improve the reliability and real-time and autonomous property of diagnosis system. In the fourth section, some RoADS ideas are adopted in the new control and diagnosis network in this chapter due to many disadvantages of the control and diagnosis networks. The design principles and plan are proposed, discussed, and analyzed in detail. Through the comparisons between old control and diagnosis networks and new one, it is concluded that new control and diagnosis network can better meet the demands of Maglev train.

In the last section, for the demand of onboard communication platform in Maglev train, a plan based on RTLinux is proposed. Firstly, the architecture design of communication platform based on RTLinux, the design and realization of real-time processing module, and non-real-time processing module are given in detail. Secondly, the driver program of RTLinux for CPCI-CAN board is given and realized. Finally, the real-time performance of communication platform of diagnosis system is analyzed by virtue of the dispatch time model and transmission delay model. The real-time simulation and analysis show that the communication platform can meet the real-time demand of actual Maglev train.

## References


1. Wang JS, Wang SY, Zheng J. Recent development of high temperature superconducting Maglev system in China [J]. IEEE Trans Appl Supercond. 2009;19(3):2142–7.
2. Yang LG. The maglev development and commercial application in China [J]. IEEE Trans Appl Supercond. 2006;16(2):1138–41.
3. Nuria DV, Alvaro S, Carles N, et al. Towards an optimized magnet-superconductor configuration in actual maglev devices [J]. IEEE Trans Appl Supercond. 2011;21(3):1469–72.
4. Jiang H, Lian J. The development and review of Maglev train in the world [J]. J China Railw Soc. 1991;13(2):88–93 (in Chinese).
5. Wu X. Maglev train [M]. Shanghai: Shanghai Science and Technology Press; 2003 (in Chinese).
6. Long ZQ, Liu SS, Li CK. The research about communication network of Maglev train based on CAN bus [J]. Electr Drive. 2001;2:44–7 (in Chinese).
7. Liu ZG. The design of diagnosis system in Maglev train [J]. IEICE Trans Inf Syst. 2005;88(12):2708–14.
8. Liu ZG, Wang Q, Tan YD. The study of maglev train control and diagnosis networks based on role automation decentralization [J]. IEICE Trans Inf Syst. 2008;91(9):2285–92.
9. Yang XH. Field bus technologies and applications [M]. Beijing: Tsinghua University Press; 2003 (in Chinese).
10. ISO 11898, Road vehicles-interchange of digital information-controller area Network for high speed communication ISO [S].
11. Ni WJ, Su AJ, Guo QY. On-board information acquisition and diagnostic system for Maglev vehicle based on CAN-Bus [J]. Electr Drive Locomot. 2006;10(6):25–8 (in Chinese).
12. Hu G, Liu ZG, Bo GF. Simulation of high-speed maglev on-board diagnosis system [J]. Urban Mass Transit. 2011;14(1):52–5 (in Chinese).
13. Salah K, Calyamb P, Buharic MI. Assessing readiness of IP networks to support desktop videoconferencing using OPNET [J]. J Netw Comput Appl. 2008;31(4):921–43.
14. Hao J, Wu JC, Guo CY. Modeling and simulation of CAN network based on OPNET [C]. In: IEEE 3rd international conference on communication software and networks. Xi'an; 2011. p. 577–81.
15. Liu ZG, Hou YC, Fu WJ. Communication simulation of on-board diagnosis network in high-speed Maglev trains [J]. J Mod Transp. 2011;19(4):240–6.
16. Kirrmann HD, Claessen U. The IEC train communication network [J]. Electr Drive Locomot. 1999;3:4–9 (in Chinese).
17. IEC61375-1. Part1: Train communication networks [S]. 1999.
18. BOSCH CAN Specification version 2.0[R]. Robert Bosch GmbH. 1993.
19. Boterenbrood H. CANopen high-level protocol for CAN-bus [M]. Amsterdam: NIKEF; Version 3.0, 2000.
20. CiA Draft Standard 301, CANopen Application layer and communication profile [S]. Version 4.02,2002.2
21. Mohammad Farsi, Karl Ratcliff. An introduction to CANopen and CANopen communication issues [J]. CANopen implementation (Digest no. 1997/384), IEE Colloquium on 6 Oct. 1997:2/1–2/6.
22. Ng JKY, Liu JWS. The performance of local area network protocols for hard real-time applications [C]. In: Proceedings of the 11th international conference on distributed computing systems. Los Alamitos: IEEE Press; 1991. p. 318–26
23. Mohammed FA, Ayad NMA, Esmat H, et al. The performance analysis of access media protocols in local networks [C]. In: Proceedings of IEEE symposium on computers and communications. Los Alamitos: IEEE Press; 1997. p. 249–53.
24. Kinji Mori. Autonomous decentralized system: concept, data field architecture and future trends [C]. In: Proceedings of ISADS 93. Kawasaki: IEEE; 1993.

25. Kinji Mori. Trend of autonomous decentralized systems [C]. In: Proceedings of the 10th IEEE international workshop on future trends of distributed computing systems. Suzhou, China, 26–28 May 2004.
26. Tan Y, Qian Q. Concept and architecture of role autonomous decentralized system [J]. China Railw Sci. 2007;28(1):99–105 (in Chinese).
27. Tan Y. Study on dynamic evolution of autonomous decentralized systems [D]. Doctor dissertation of Southwest Jiaotong University; 2007 (in Chinese).
28. Bbarabanov M. A Linux based real time operating system [D]. New Mexico: New Mexico Institute of Technology; 1997.
29. Zhigang Liu, Zhiwei Han, Quanwei Peng, Zhiguo Zhou. Design and realization of on board communication platform for high-speed Maglev train based on RTLinux [J]. J Southwest Jiaotong Univ. 2010;45(6):830–5.

# Chapter 6
# The Position and Speed Detection Technology Based on Loop Cable for Low-Speed Maglev Train

## 6.1 Introduction

The method of linear asynchronous traction is adopted for the low-speed Maglev train. The primary winding of linear motor is installed at the bottom of the train bogie, while the rail is the secondary winding. The ability of effectively and accurately obtaining speed information would relate to the traction efficiency of the motor. Since different Maglev trains have different structures, and the position detection methods are distinguishing, the mode of low-speed maglev trains that use the asynchronous traction way with short stators is selected [1]. The basic principles of the position and speed detection method based on the loop cable are described as follows.

The characteristics of Maglev trains determine that the speed and location detection techniques of wheel-track trains cannot be adopted directly. At present, some common methods for the speed and location detection of Maglev trains are listed as follows:

1. Position and speed detection based on sleeper tally

A group of four eddy-current type switch sensors are fixed under the train with a distance of 0.3 m. When the train goes over the sleeper, there will be four pulse signals. As these four pulse signals are added, the position sequence pulses can be obtained. Therefore, the speed and relative orientation can be calculated by these position sequence pulses.



Z. Liu et al., *Maglev Trains*, Springer Tracts in Mechanical Engineering,
DOI 10.1007/978-3-662-45673-6_6

This method can fulfill the requirement of traction and braking. With the method, the speed and relative orientation can be easy to obtained, but the disadvantages are that the sleeper must be laid equidistantly and its precision is poor.

2. Position and speed detection based on Doppler radar

The Doppler radar method is based on the "Doppler effect" of microwave transmission. The radar's antenna ends the electromagnetic wave along the railway and gets the reflected wave. Due to the relative movement between the train and the rail, there is great distinction between the transmitted wave and the received wave. As the distinction is related with the speed of train, the speed can be calculated.

In practice, the radar antenna must be fixed in such a correct angle that can make the pitch angle and drift angle coincide with the design parameters. Errors often appear in the rain and snow, so the detection precision may decrease.

3. Position and speed detection based on long stators

The long stator synchronous machine is used in German Maglev train. Its speed and location detection is based on the long stator. Since the long stator has a very accurate geometrical size, it is economical and accurate to make use of it. The method can adopt a special sensor to detect the change of inductance. As the sensor's coil faces the tooth surface, the inductance is different from that when the sensor's coil faces the groove surface.

The speed measurement based on long stators is one kind of passive methods. This system is economical and reliable and needs little maintenance. In addition, the signal processing is easy. However, it requires the special long stators with the structure of tooth space. TR-series high-speed Maglev trains are equipped with this method [2].

4. Position and speed detection based on the inductive loop cable

The position and speed detection based on the inductive loop cable adopts the electromagnetic induction principle to detect the position and speed of the train. To start with, the loop cable with the fixed shape is laid along the rails, and a high-frequency alternating current, which is also called the carrier current, is injected into the loop cable. Meanwhile, an alternating magnetic field with the same frequency is produced. As the current of the adjacent loop-cable rings goes in the opposite direction, the adjacent rings would produce the magnetic field with the opposite direction. Then, the receiving antenna composed of several sets of coils is installed above the loop cable at the bottom of the train. According to the Faraday's law of electromagnetic induction, an inductive electromotive force is generated between both ends of the coils. When the train runs along the rails, the position and speed information could be determined by the amplitude and phase change of the inductive electromotive force. HSST-series low-speed Maglev train applies this detection method [3–5].

In the following sections, the principles of the speed and proposed position measurement method based on the inductive looped cable will be introduced in detail.

## 6.2 Position and Speed Detection Based on the XOR Pulse

### *6.2.1 The Electromagnetic Field Analysis of Loop Cable*

First, the magnetic flux density of limited long current-carrying wire is calculated and the point $P$ is chosen arbitrarily, as shown in Fig. 6.1. As $Id\overline{l} = Idz\overline{e}_y$ and $\overline{R} = \rho\overline{e}_\rho - y\overline{e}_y$, it could be obtained as follows:

$$Id\overline{l} \times \overline{R} = I\rho dy\overline{e}_\phi \tag{6.1}$$

According to the Biot–Savart law, the magnetic flux density of the point $P$ in Fig. 6.1 is calculated as follows:

$$\begin{aligned} \overline{B} &= \frac{\mu_0 I\rho}{4\pi}\int_a^b \frac{dz}{[\rho^2 + b^2]^{3/2}}\overline{e}_\phi \\ &= \frac{\mu_0 I}{4\pi\rho}\left[\frac{b}{\sqrt{\rho^2+b^2}} - \frac{a}{\sqrt{\rho^2+a^2}}\right]\overline{e}_\phi \end{aligned} \tag{6.2}$$

Second, the flux density at any point $P$ outside the square coil with current $I$ is calculated, which is shown in Fig. 6.2.

$$\vec{B_P} = \vec{B_{AB}} + \vec{B_{BC}} + \vec{B_{CD}} + \vec{B_{DA}} \tag{6.3}$$

Therein, the magnetic flux density generated by all square current-carrying coils is listed as follows:

$$\vec{B_{AB}} = \frac{\mu_0 I}{4\pi\rho_1}\left[\frac{l/2 + y}{\sqrt{\rho_1^2 + (l/2 + y)^2}} + \frac{l/2 - y}{\sqrt{\rho_1^2 + (l/2 - y)^2}}\right]\vec{e}_{\phi 1} \tag{6.4}$$

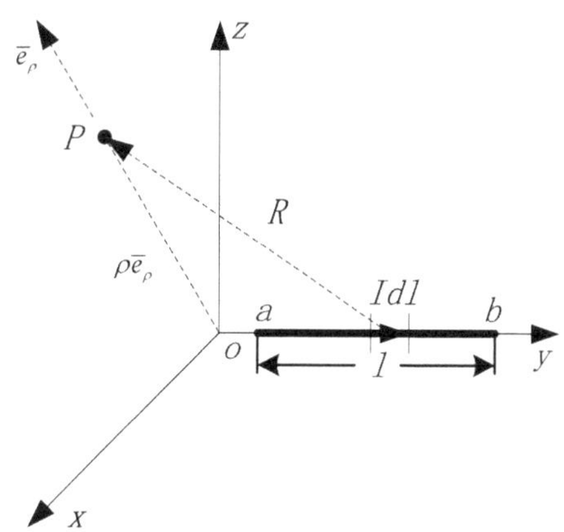


**Fig. 6.1** The magnetic flux density of finite current-carrying conductor

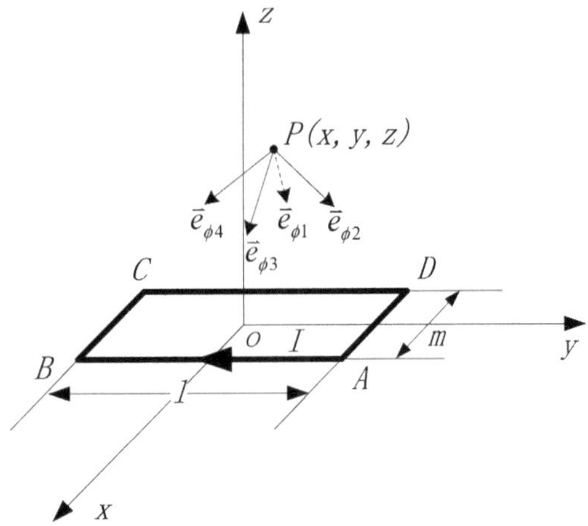


**Fig. 6.2** The magnetic flux density of a square current-carrying coil

$$\vec{B}_{BC} = \frac{\mu_0 I}{4\pi\rho_2}\left[\frac{m/2+x}{\sqrt{\rho_2^2+(m/2+x)^2}}+\frac{m/2-x}{\sqrt{\rho_2^2+(m/2-x)^2}}\right]\bar{e}_{\phi 2} \tag{6.5}$$

$$\vec{B}_{CD} = \frac{\mu_0 I}{4\pi\rho_3}\left[\frac{l/2-y}{\sqrt{\rho_3^2+(l/2-y)^2}}+\frac{l/2+y}{\sqrt{\rho_3^2+(l/2+y)^2}}\right]\vec{e}_{\phi 3} \tag{6.6}$$

$$\vec{B}_{DA} = \frac{\mu_0 I}{4\pi\rho_4}\left[\frac{m/2-x}{\sqrt{\rho_4^2+(m/2-x)^2}}+\frac{m/2+x}{\sqrt{\rho_4^2+(m/2+x)^2}}\right]\bar{e}_{\phi 4} \tag{6.7}$$

where $\rho_1 = \sqrt{z^2+\left(\frac{m}{2}-x\right)^2}$, $\rho_2 = \sqrt{z^2+\left(\frac{l}{2}+y\right)^2}$, $\rho_3 = \sqrt{z^2+\left(\frac{m}{2}+x\right)^2}$ $\rho_4 = \sqrt{z^2+\left(\frac{l}{2}-y\right)^2}$.

Figure 6.3 shows the inductive loop cable laid on railway, and the loop-cable coil plane is placed in *xoy* plane of the space rectangular coordinate system, composed of several unit rings with back-to-back connection. Each unit ring is 30 cm in length and 10 cm in width.

Since lacuna ($l$) between rings is much less than $L$, the loop cable can be equivalent to several rectangular current-carrying coils in a proper order, and a high-frequency alternating current is injected into each rectangular coil, which is shown in Fig. 6.4.

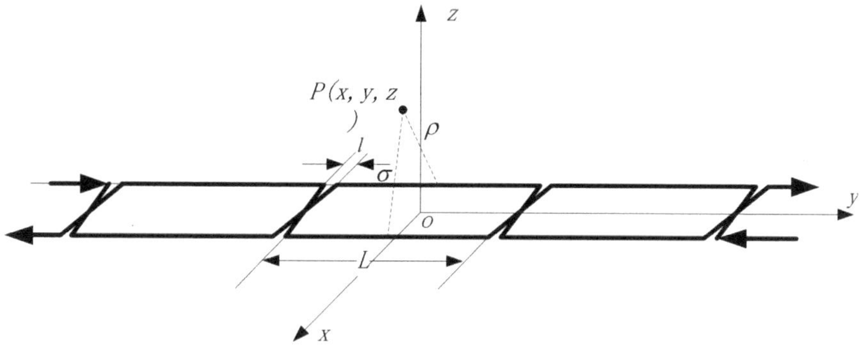


**Fig. 6.3** Model of current-carrying inductive loop cable

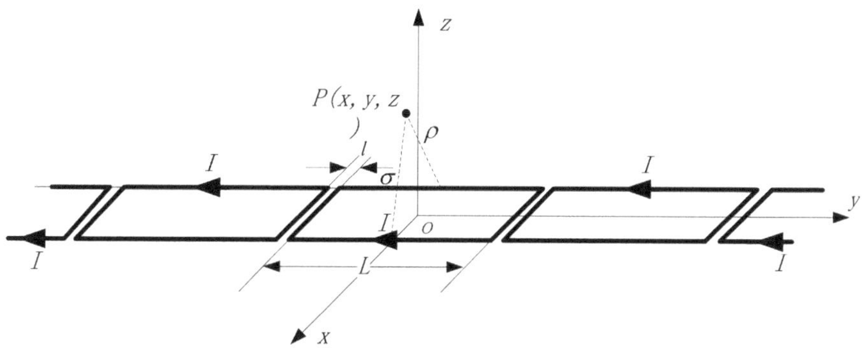


**Fig. 6.4** Equivalent model of current-carrying inductive loop cable

The magnetic flux density at any point $P$ located outside the loop cable is superposition of that generated by all rectangular current-carrying coils. Due to periodical distribution along $y$ axis of inductive loop cable, the magnetic flux density is destined to present the periodical change along $y$ axis.

The 10 cm height away from railway loop cable is chosen as the detection surface of magnetic field. According to Formulas (6.4), (6.5), (6.6), and (6.7), in case 1 a constant current is injected into the loop cable; the spatial distribution curve of the magnetic flux density in three directions will be obtained with finite element simulation software (Ansoft), as shown in Figs. 6.5, 6.6, and 6.7. The component of magnetic flux density in each direction is chosen as a measured physical variant, and the relative position of the train and speed change can be calculated by measuring it [6, 7].

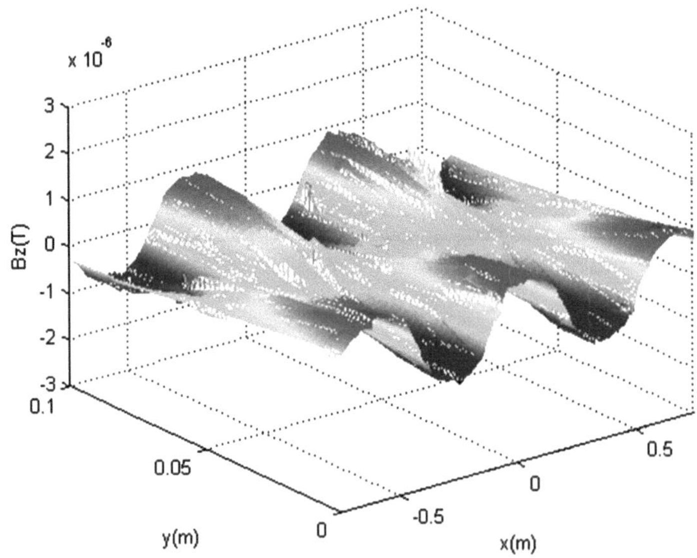


**Fig. 6.5** Illustration of the variance curve of the magnetic flux density along *x* direction

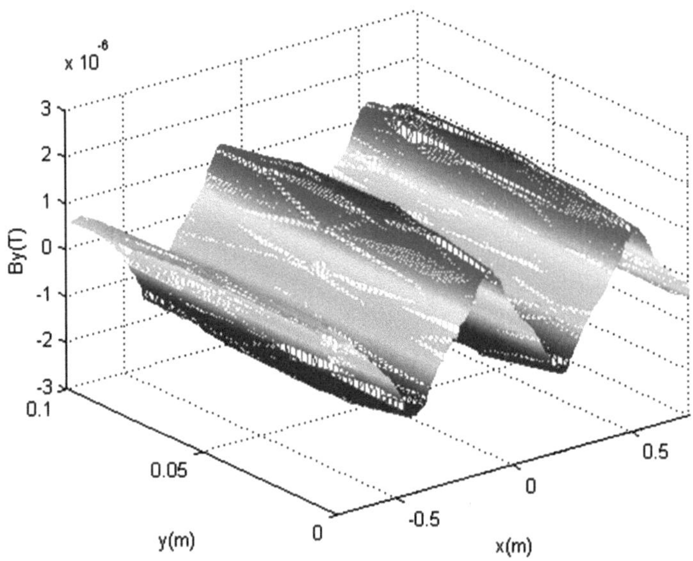


**Fig. 6.6** Illustration of the variance curve of the magnetic flux density along *y* direction

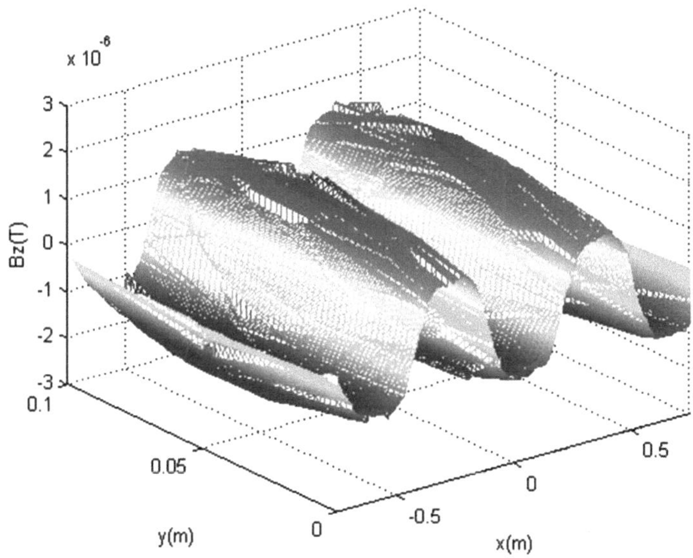


**Fig. 6.7** Illustration of the variance curve of the magnetic flux density along *z* direction

## 6.2.2 Detection of Component of Magnetic Flux Density

In Sect. 6.2.1, variance curves of the magnetic flux density are obtained by injecting a constant current into the inductive loop cable. However, it is a high-frequency alternating current that is actually injected into loop cable, and correspondingly the alternating field is generated. It can be treated as the magnetic quasi-static field on condition that vehicle-mounted receiving coil cares about the amplitude of the magnetic flux density, and the distance between that and field domain is much less than the wavelength. The absolute values of curves shown in Figs. 6.5, 6.6, and 6.7 can represent the amplitude of magnetic flux density generated by the inductive loop cable.

The vehicle-mounted receiving coil is located above the railway inductive loop cable, and its different installation manners decide its different components of magnetic flux density. Generally speaking, it can be divided into the three following detection modes [8, 9]:

1. The magnetic field detection along *x* axis, shown in Fig. 6.8. When the receiving coil plane is parallel to *yoz* plane of coordinate axis, receiving coil will couple with magnetic field along *x* direction.

   Variance curves of the magnetic flux density of inductive loop cable along x direction are shown in Fig. 6.5. When receiving coil is placed right above the inductive loop-cable edge, namely, position *B* in Fig. 6.8, the magnetic flux density becomes the maximum. When the receiving coil is placed in *yoz* plane, namely, position *A* in Fig. 6.8, its output signal is zero since the coupling

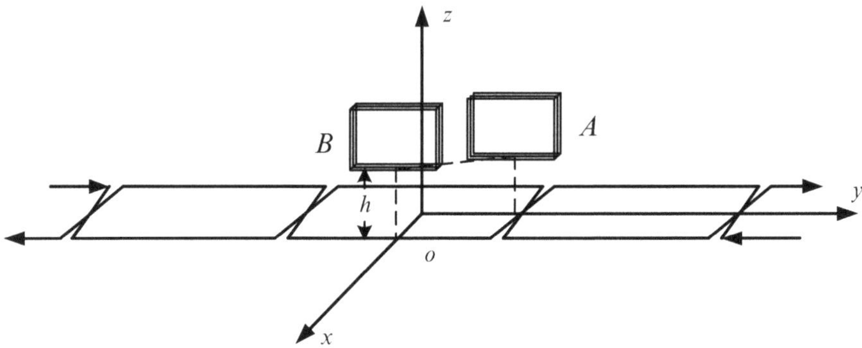


**Fig. 6.8** Receiving coil plane parallel to *yoz* plane of coordinate axis

magnetic flux of coil is zero. When the coil continues to move in a constant speed, the receiving coil can generate inductive electromotive force reflecting the relative position of train and the periodical variance of amplitude envelope. Hence, the receiving antenna shall be placed right above one side of loop cable if this installation mode is chosen.

2. Magnetic field detection along *y* axis. When the receiving coil plane is parallel to *xoz* plane of coordinate axis, the receiving coil will couple with the magnetic field along *y* direction. The illustration of installing receiving coil is shown in Fig. 6.9. It can be seen from variance curves of the magnetic flux density of the inductive loop cable along *y* direction in Fig. 6.6; when the centerline of the receiving coil is placed above cross-point of the inductive loop cable, the amplitude of coupling magnetic flux is the maximum. When the centerline of receiving coil is placed right above the centerline of unit ring of the inductive loop cable, the coupling magnetic flux becomes close to zero. When the coil moves in a constant speed along *y* axis, the receiving coil can generate inductive electromotive forces that can reflect the relative position of train and periodical variance of amplitude envelope.
3. Magnetic field detection along *z* axis. When the receiving coil plane is parallel to *xoy* plane of the coordinate axis, the receiving coil will couple with magnetic field along *z* direction. Illustration of installing the receiving coil is shown in Fig. 6.10.

It can be concluded from variance curves of the magnetic flux density of the inductive loop cable along *z* direction in Fig. 6.7 that the component amplitude in this direction presents periodical variance above the loop cable. When the centerline of the receiving coil is placed above that of unit ring of the inductive loop cable, the amplitude of its coupling magnetic flux is the maximum. When the receiving coil is placed above the cross-point between unit rings, the amplitude of its coupling magnetic flux is close to zero. When the coil moves in a constant speed, it can generate inductive electromotive force reflecting the train position.

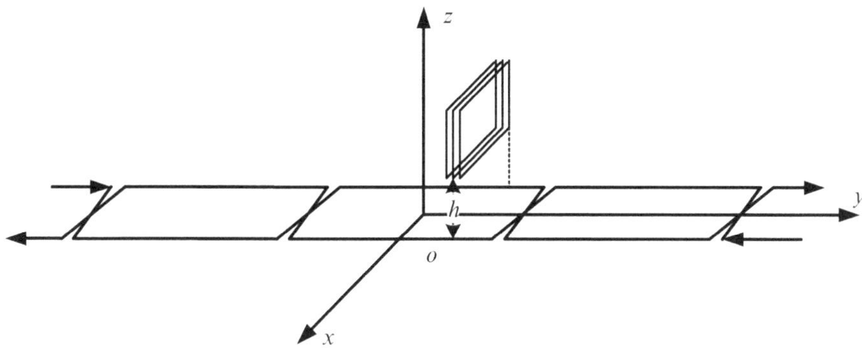


Fig. 6.9 Receiving coil plane perpendicular to *yoz* plane of coordinate axis

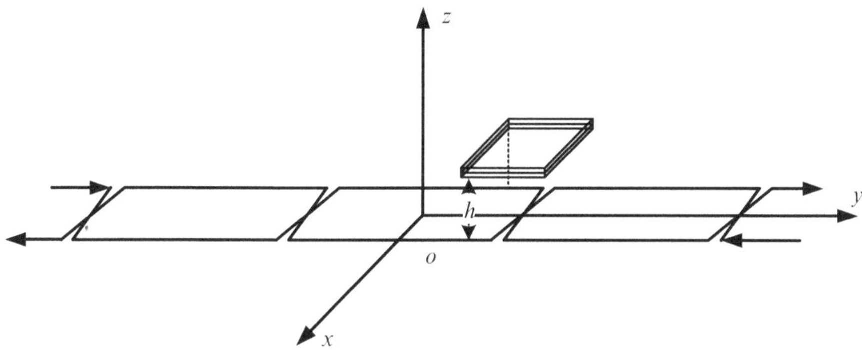


Fig. 6.10 Receiving coil plane perpendicular to *xoy* plane of coordinate axis

The position and speed detection system using the inductive loop cable demands that the receiving coil should be sensitive to directional variance of the measured component [10]. Through comprehensive comparison of curve variance of the magnetic flux density along $x$, $y$, and $z$ axis, it can be shown that $x$ direction component amplitude of the magnetic flux density of the detected coil is relatively small, which is unsuitable as a measured component. Theoretically, the phase of the $y$ direction magnetic flux density component sharply changes in the right middle of the loop-cable unit ring, and that along $z$ direction sharply changes in the cross-point of the inductive loop-cable unit ring. Such a difference will hardly have a fundamental effect on relative position detection of the position and speed detection system. However, in order to improve the ability of dynamic response of the detected coil, the larger the variance rate of magnetic flux density where the magnetic field direction sharply changes is, the more helpful for the rapid response to receive signal for the receiving coil and the more accurate for the position signal detection. By contrast of the magnetic flux density in $y$ and $z$ direction components, the change rate of the magnetic flux density along $z$ direction near the cross-point is bigger than that along $y$ direction. Therefore, the magnetic flux density along $z$ direction is more suitable as a detected component. When the magnetic flux density along $z$ direction is chosen as the detected component of vehicle-mounted receiving antenna, a fixed form of the receiving antenna can also be determined regardless of the detected coil plane parallel to the plane of the inductive loop cable, as shown in Fig. 6.10.

## 6.3 Design of Receiving Coil

### *6.3.1 Selection of Resonance Circuit*

To guarantee the receiving antenna to effectively receive the specific frequency signal sent by the ground loop cable and to restrain other frequency signal at the same time, the resonance loop is proposed as shown in Fig. 6.11. It owns frequency-selection characteristics and can be used as band-pass filter. Resonance characteristics of the resonance loop are related to its quality factor. Namely, the larger the quality factor is, the better the frequency-selection characteristics will be.

The quality factor of an $LC$ resonance loop is defined by the specific value of the resonance circuit loop characteristic impedance below:

$$Q = \frac{\omega_0 L}{R_0} \tag{6.8}$$

Therein, $R_0$ is the internal impedance of coil, $\omega_0$ is the resonance angular frequency, and $L$ is the coil inductance. As the receiving antenna, selection of the resonance frequency depends on the frequency of the transmission signal, which refers to the carrier frequency of the ground inductive loop table in this section. Considering that loss of higher-frequency signal will more easily increase in the long-distance transportation due to skin effect, so the signal frequency of driving inductive loop cable should not be too large and the selected signal frequency is 100 kHz actually.

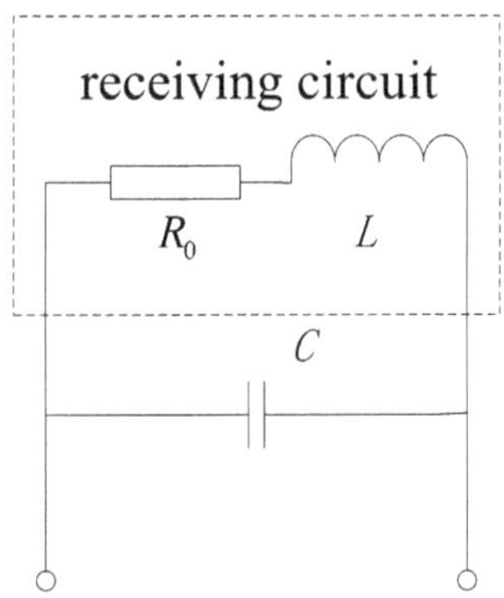


**Fig. 6.11** Resonance circuit of receiving circuit

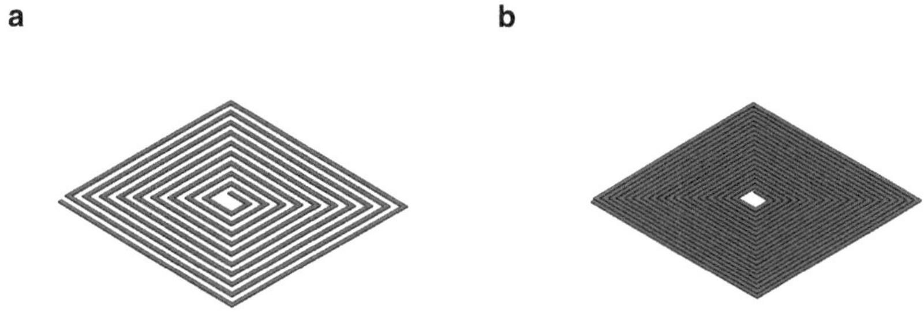


**Fig. 6.12** Coils made of printed board

### 6.3.2 Design of Coil Structure

The coil as shown in Fig. 6.12 is considered to be used. *A* and *B* coils are made of printed board, which shall be intertwined around the center and be located in the same plane with size of 10 cm × 10 cm. And the intertwining gap between coils shall be uniform. The number of turns in *A* coil is 20 and in *B* 40.

Both *A* and *B* coils are made of printed board with simple process and good consistency, and the distribution parameter of the same coils is relatively stable. Among isotropy media, in case that the magnetic field is generated by some current loop, the inductance of this coil loop is the specific value of the magnetic flux passing through the restricted area of this current loop and current of the current loop. Intuitively, *A* and *B* coils are intertwined in a coil nest. The magnetic field generated by external coil loop cancels with that by the internal coil loop, causing the inductance of coils made in this manner to be relatively small and the quality factor to be also not large, as shown in Fig. 6.13. In addition, when the number of coil turns increases, the quality factor will also increase. When driven by signal with a frequency of 100 kHz, the measured quality factors are 12.08 and 14.52, respectively.

In order to overcome the issue of the low-quality factor due to the relatively small inductance of *A* and *B* coils, coils in intertwining manner shown in Fig. 6.14 can be adopted. And the size of *C* and *D* coils is 10 cm × 10 cm, and an enameled copper wire with diameter of 0.5 mm is used to intertwine. The number of turns of *C* and *D* coils is 20 and 40 respectively. In Fig. 6.15, it can be seen from function curve of quality factor that the quality factors of *C* and *D* coils are 56.71 and 59.27, respectively, for a drive signal of 100 kHz. Even though the number of turns in *C* and *D* coils is the same as that of *A* and *B*, the quality factors corresponding to the same frequency are much larger. Since the number of turns in *D* coil is twice as many as that of *C* coil, its quality factor is much larger in the relatively low-frequency range. By respectively comparing *A* with *B* coils, and *C* with *D* coils, it can be found that on the premise of the same distribution of the magnetic flux density, the more the number of turns is, the larger the magnetic induced electromotive force is, and the

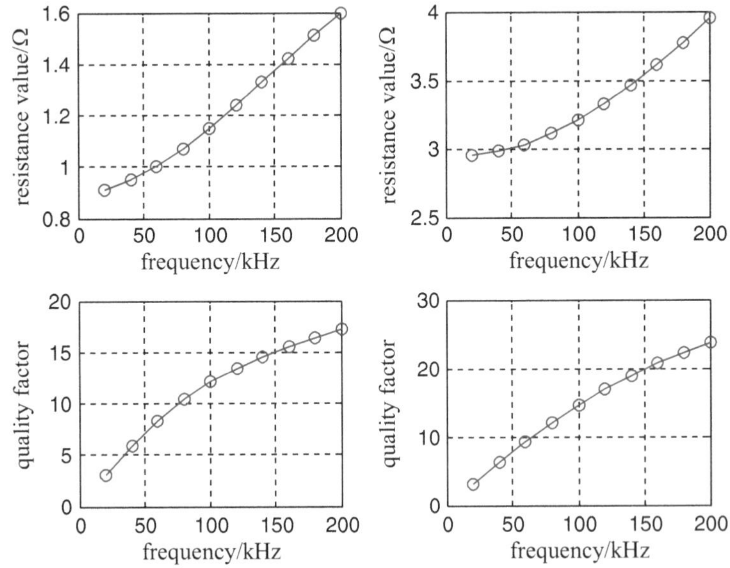


**Fig. 6.13** AC internal impedance and quality factor of *A* and *B* coils

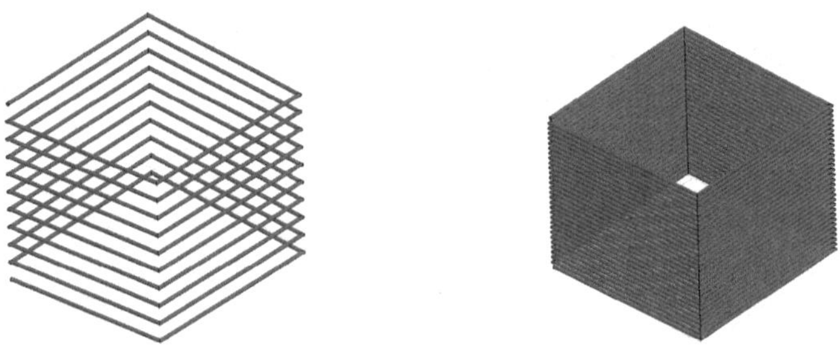

**Fig. 6.14** Coil made in enameled wire intertwining manner

stronger the receiving signal strength is. Hence, properly increasing number of turns of coils is beneficial to signal reception.

In addition, Formula (6.8) illustrates that the quality factor of a coil is related to its inductance, its resonance angular frequency, and its AC internal impedance. Under *AC* condition, due to occurrence of skin effect, the distribution of current and electromagnetic field inside conductor is centralized on the surface of conductor. After the depth is greater than several depths of penetration, they are all approximate to zero. Even though the cross-sectional area of the conductor is relatively large, most has not been used, and its actual current-carrying cross-sectional area will decrease. Therefore, under *AC* condition, the internal self-inductance of the conductor will differ from that under *DC* condition, and its resistance will increase with frequency. As a result, the function of quality factor is nonlinear, which can

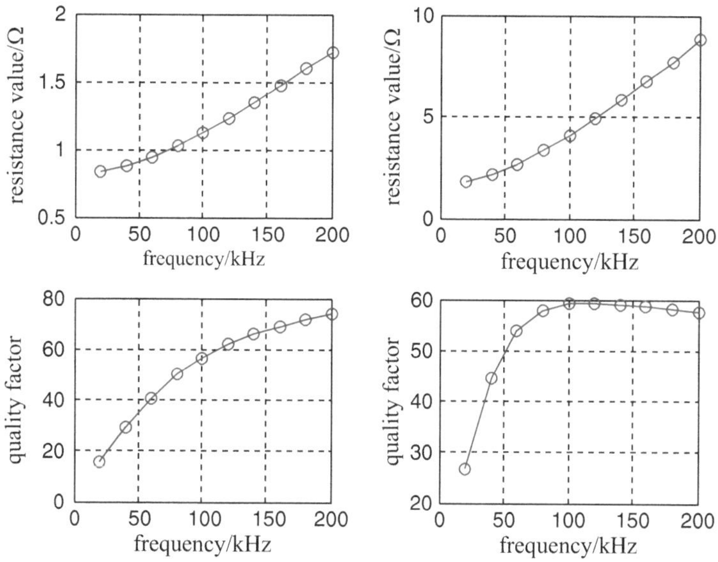


**Fig. 6.15** AC internal impedance and quality factor of *C* and *D* coils

be clearly seen from the function curve of quality factor and internal resistance of *D* coil as shown in Fig. 6.15. Under the increasing-frequency condition, especially after more than 100 kHz, the quality factor does not change greatly.

After comprehensively considering the quality factor of the coil, the strength of the receiving signal and other factors, in ideal conditions, *D* coil shall be selected as the basic structure of the receiving antenna.

## 6.4 Signal Processing Methods

The position and speed detection system based on the loop cable for the low-speed Maglev train lays and fixes the inductive loop cable on the surface of the sleeper [10–12]. And the antenna bracket is installed inside train bogies; meanwhile, the receiving antenna is lifted right above the inductive loop cable through the bracket. The unit ring of the inductive loop cable is 30 cm in length and 10 cm in width.

The receiving antenna is composed of several sets of coils. Different configuration formation and coil numbers shall be used according to detection precision and testing principal. Based on the analysis in Sect. 6.2, it can be seen as follows: when a high-frequency alternating current is injected into the inductive loop cable laid on railway, coils of the receiving antenna will generate induced electromotive force through magnetic flux coupling and induced electromotive force will present different amplitude and phase characteristics according to relevant position between the inductive loop cable and the coil. Therefore, two different signal processing methods based on amplitude or phase can be used [8].

## 6.4.1 Signal Processing Method Based on Amplitude Detection

### 6.4.1.1 Extraction of Position Pulse

When the train runs, the receiving antenna moves towards the inductive loop cable, and the receiving coil inside the antenna generates induced electromotive force. When the coil center is far away from the cross-point of the loop cable, projection of the receiving coil inside the inductive loop cable is in the same ring, and the amplitude of inductive electromotive force at its two ends is greater than zero. When the coil gradually approaches to the cross-point of loop cable, the projected area of receiving antennas generated inside two adjacent rings gradually are equal each other, and according to the analysis in the last chapter, its induced electromotive force is close to zero as shown in Fig. 6.16. By extracting the amplitude characteristics of the receiving coil-induced electromotive force at the cross-point, the corresponding amplitude detection circuit and threshold comparison circuit shall be designed, and the extraction of the pulse can be realized.

In order to effectively improve anti-interference capacity, on the basis of the original receiving coil, added with a coil of the same size and the same number of turns as the receiving coil, a differential mode receiving coil is composed in manner of in-phase connection as shown in Fig. 6.17.

Each time the receiving coil passes a cross-point, the threshold comparison circuit will generate a narrow pulse, and the positional precision is 30 cm, the length of unit ring. Installation mode of multi-coils can be adopted to increase precision. Figure 6.18 shows the installation mode of triple coils, the center distance of each coil is one third of unit ring length, and then a narrow pulse of 10 cm is generated. The signal processing method is shown in Fig. 6.19.

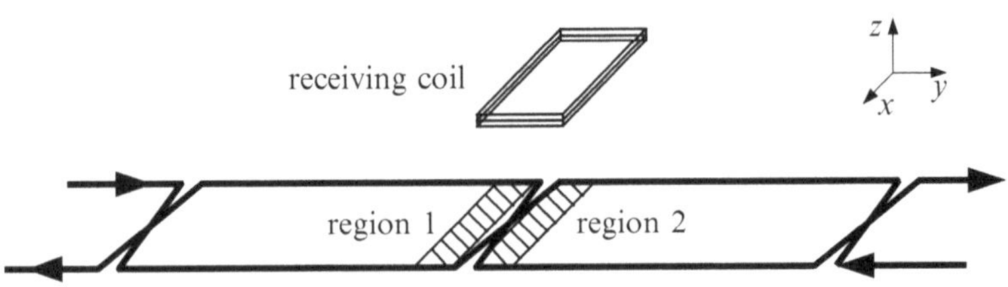


**Fig. 6.16** Projection area of receiving coil on inductive loop cable

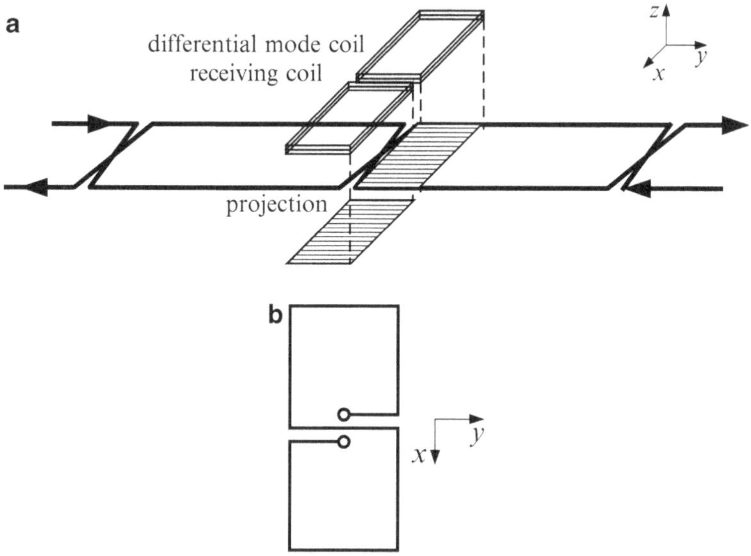


**Fig. 6.17** Differential mode receiving coil. (**a**) Illustration of differential mode receiving coil. (**b**) Illustration of differential mode receiving coil connection

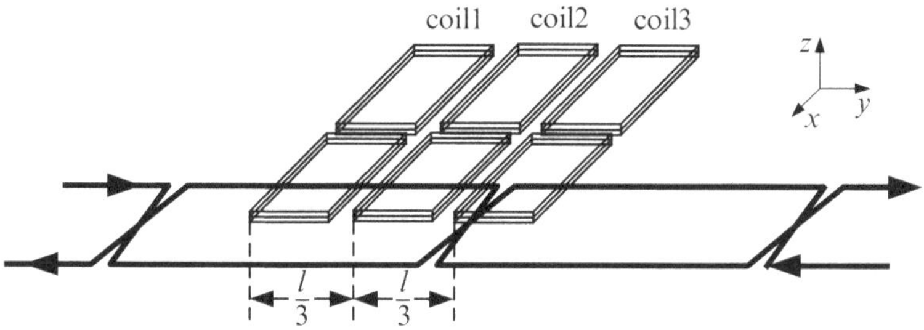


**Fig. 6.18** Illustration of triple coils configuration

#### 6.4.1.2 Discretion of Direction

According to the sequence of coil 1, 2, and 3 passing the cross-point, the moving direction of the train can be distinguished. Detected cross-point pulses of three coils are marked as pulse $I$ ($i = 1, 2, 3$) in sequence, and the rule of direction discretion shown in Table 6.1 can be obtained.

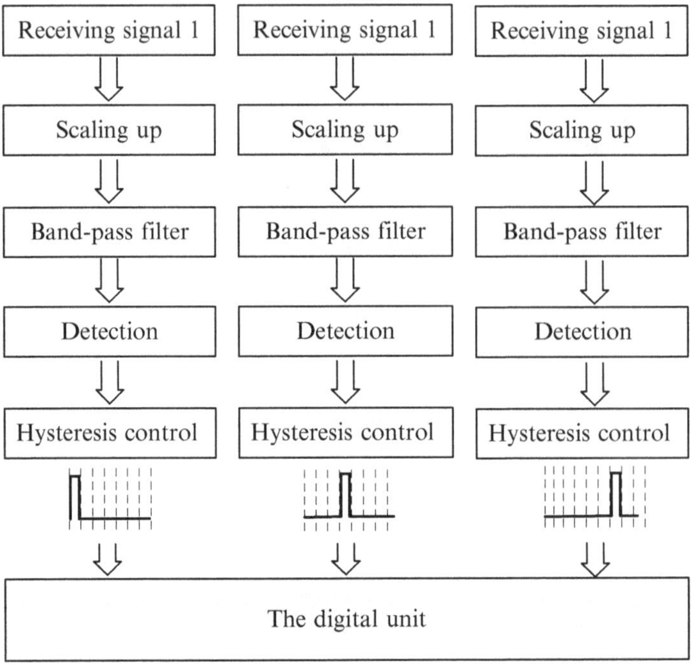


**Fig. 6.19** Processing flow of triple coil receiving signal

**Table 6.1** Direction discretion rules

| Pulse $N$ | Pulse $N+1$ | Direction |
|---|---|---|
| 1 | 2 | Forward |
| 2 | 3 | Forward |
| 3 | 1 | Forward |
| 3 | 2 | Reverse |
| 2 | 1 | Reverse |
| 1 | 3 | Reverse |

## *6.4.2 Signal Processing Method Based on Phase Detection*

### 6.4.2.1 Extraction of Position Square Wave of Single Antenna Pair

Since the inductive loop cable adopts cross-structure, the current direction injected into two adjacent unit rings is always converse, and two adjacent rings always generate alternating magnetic fields with converse direction.

As shown in Fig. 6.20, a pair of coils is used as a detection unit, and the center distance of two coils is a half of the unit ring length. When centers of two receiving coils from one detection unit are located above the same unit ring, which is shown in Fig. 6.21, the receiving coil will generate in-phase induced electromotive force. With the detection unit moving along the railway inductive loop table, the

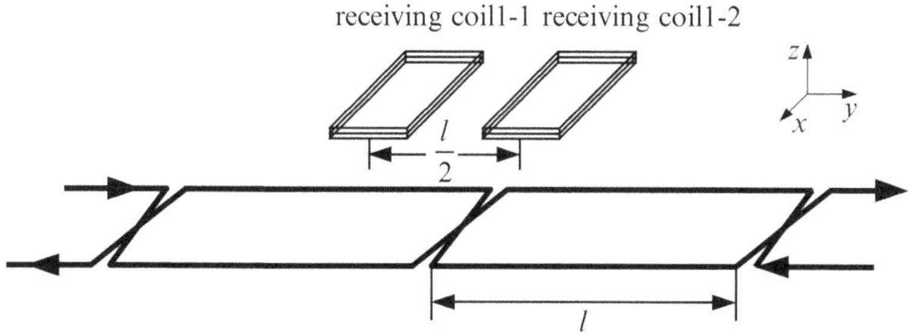


**Fig. 6.20** Phase detection coil pair

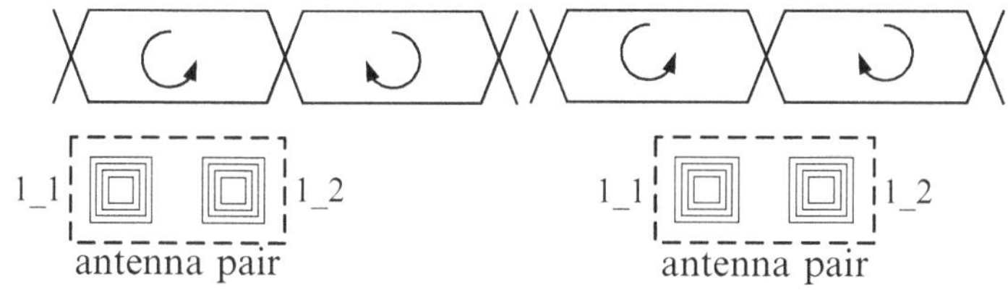


**Fig. 6.21** Illustration of relation of antenna pair and loop-cable position

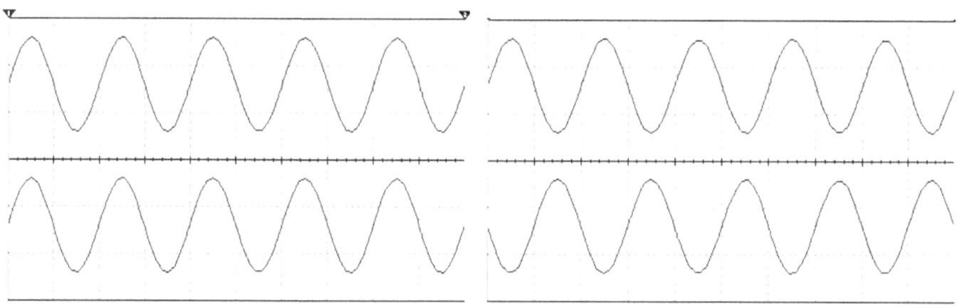

**Fig. 6.22** Receiving signals of two coils under two position conditions

phase of induced electromotive force of two coils will periodically present two circumstances, namely, "in-phase" and "anti-phase." Under the ideal condition, in-phase distance is the same as anti-phase distance, which is $l/2$. Figure 6.22 respectively shows oscilloscope screenshot of the receiving signal under in-phase and anti-phase condition.

The inverter-fed oscilloscope screenshot waveform shown in Fig. 6.23a, b can be obtained after the receiving signal of two coils processes voltage following linear amplification, band-pass filtering, and zero-crossing comparison treatment. Then the analog signal can be transferred into the digital signal. The square signal reflecting the relative position of train can be obtained after these two logical square signals further process exclusive or treatment; exclusive or output in Fig. 6.23a is logic "0" and in Fig. 6.23b "1." When the detection unit moves along the inductive loop cable, an exclusive or square signal representing phase change can be generated as

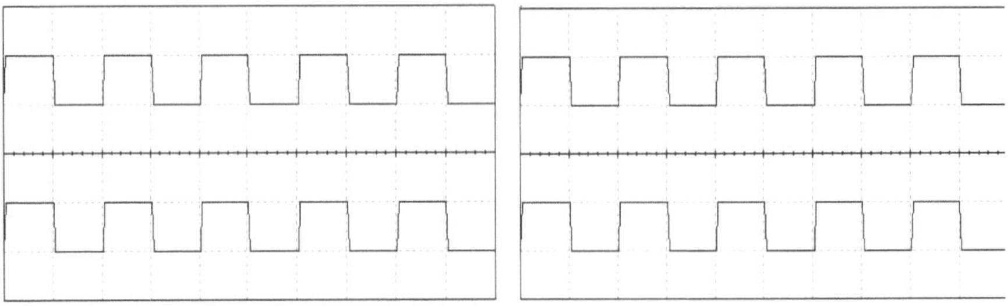

**Fig. 6.23** Oscilloscope screenshot signal after shape-correction treatment of signal in Fig. 6.22

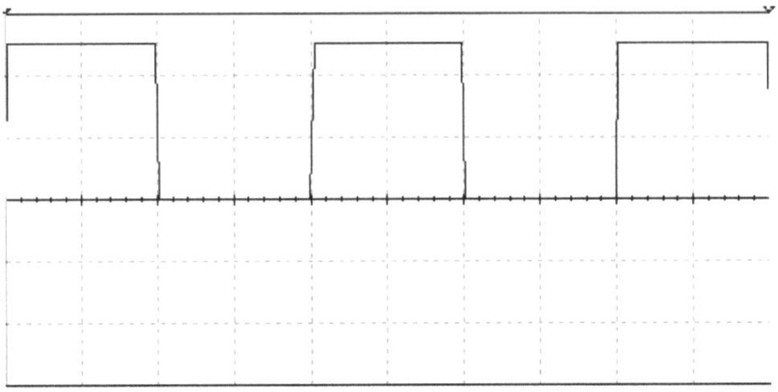

**Fig. 6.24** Oscilloscope screenshot of unit ring detecting exclusive or square signal

**Fig. 6.25** Illustration of three antenna pairs

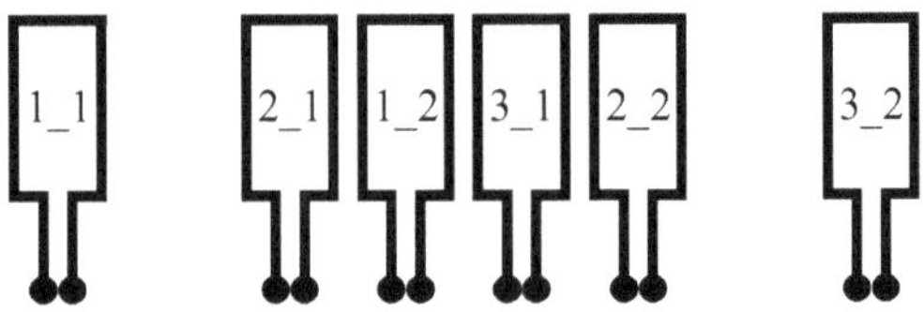


shown in Fig. 6.24. By counting rising edge or falling edge of the square wave, the relative position of train can be obtained, and the speed of the moving train can be calculated.

#### 6.4.2.2 Extraction of Position Square Wave of Multi-Antenna Pairs

With the detection method based on the amplitude by using multi-coils, the signal processing scheme based on phase detection, the antenna pair with certain regular array can also be added to enhance positioning precision as shown in Fig. 6.25. Therein, the coil 1_1 and coil 1_2 consist of an antenna pair, the coil 2_1 and coil 2_2 consist of an antenna pair, and coil 3_1 and coil 3_2 consist of an antenna pair. To strengthen the system anti-interference performance and to improve the signal to noise ratio (SNR), a coil with the same size and number of turns shall be added

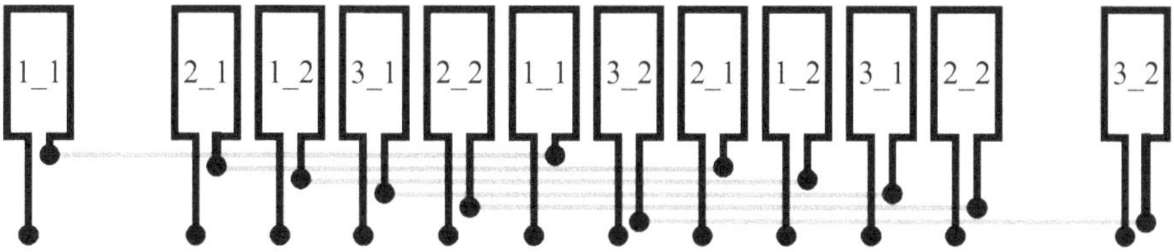


**Fig. 6.26** Illustration of antenna pairs with differential mode coil structure

to each coil to form a differential mode receiving coil. The center distance of two coils is the length of a unit ring. Thereby two coils can always be placed above incongruous unit ring to realize differential mode enhancement of effective signal and common mode rejection of interference signal generated by the inductive loop cable. The specific position configuration is shown in Fig. 6.26.

Since the center distance of each antenna pair is one third of the unit ring length, three square signals of whose phase difference lags 120° successively are generated, recorded as square wave $I$ ($i = 1, 2, 3$), and the processing flow before generating square signal is shown in Fig. 6.27.

The square signal with three times of precision after treatment can be obtained after three square signals with 120° phase difference undergo logical conversion. The logical conversion manner is shown in Fig. 6.28, of which the corresponding logical relationship expression is listed as follows:

$$Y = \overline{P1} \bullet P2 \bullet P3 + P1 \bullet \overline{P2} \bullet P3 + P1 \bullet P2 \bullet \overline{P3} \qquad (6.9)$$

Therein, $Pi$ ($i = 1, 2, 3$) corresponds to three input square signals, $Y$ is the relative position square signal after segmentation, and the signal processing result is shown in Fig. 6.29.

#### 6.4.2.3 Direction Discretion

The moving direction of the train can be judged by reading the electrical level of another square signal when the level of one square signal changes with time. According to Fig. 6.29, the following direction discretion rules can easily be obtained below (Table 6.2).

## 6.4.3 *Comparison of Two Signal Processing Methods*

The signal processing method based on amplitude detection is featured with a simple coil structure, and on the premise of guaranteed precision, the number of necessary coils is a half of that for the signal processing method based on phase. However,

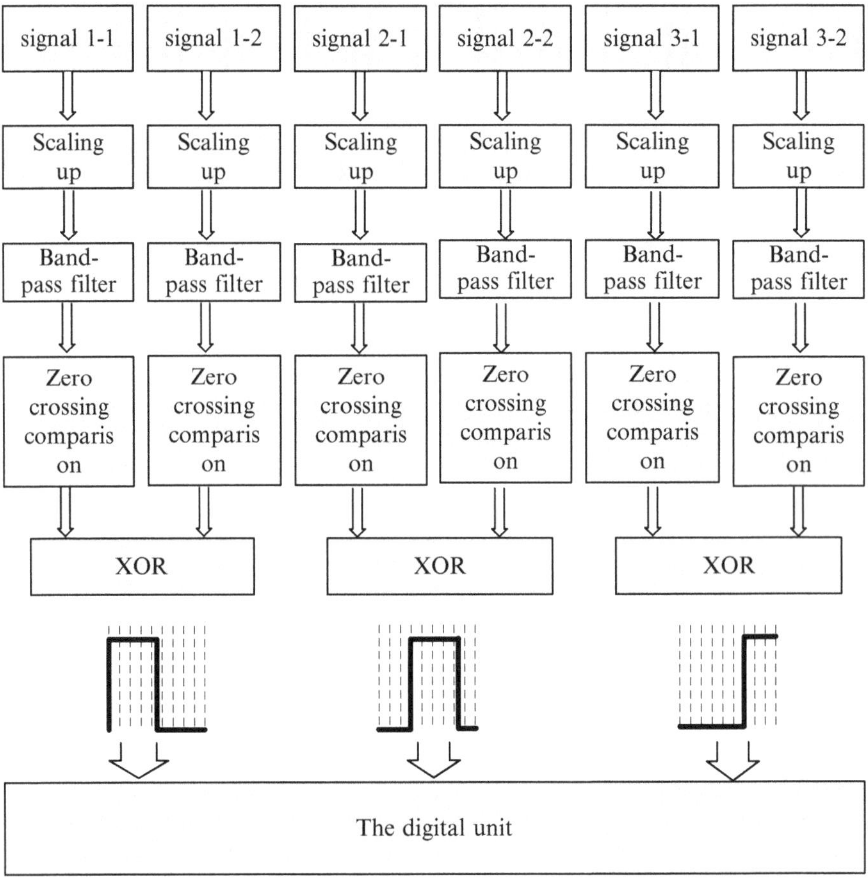


**Fig. 6.27** Illustration of three antenna pair structure

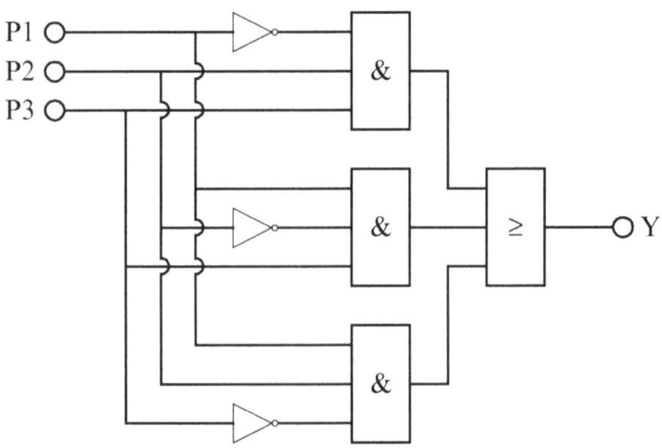


**Fig. 6.28** Logical relationship figure of three square wave segmentation treatment

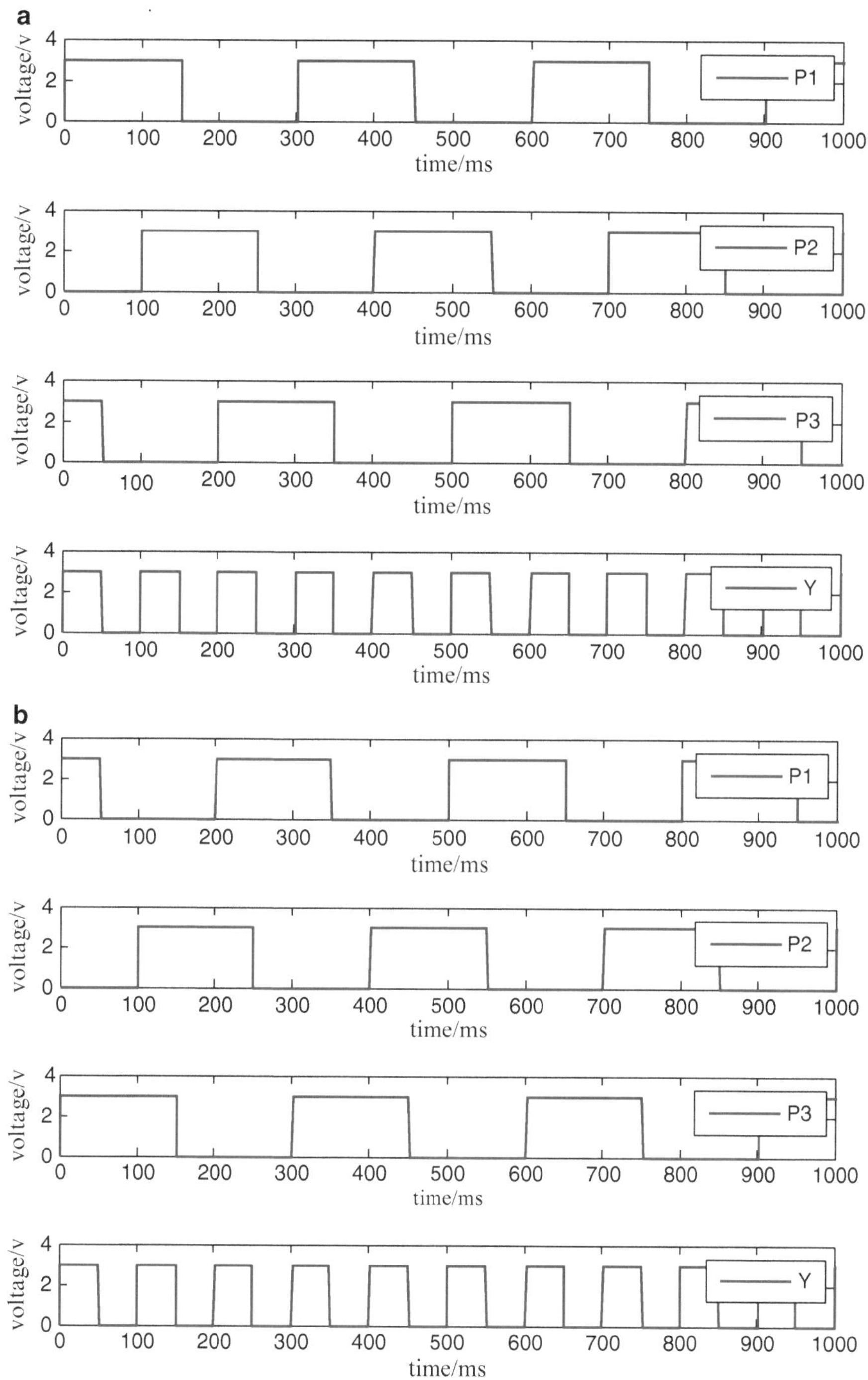


**Fig. 6.29** Three output square waves and exclusive or square wave: (**a**) forward direction and (**b**) reverse direction

**Table 6.2** Direction discretion rules

| Square signal | Way of change | P1 electrical level | P2 electrical level | P3 electrical level | Direction |
|---|---|---|---|---|---|
| P1 | Rising edge | | Low level | High level | Forward |
| P1 | Falling edge | | High level | Low level | Forward |
| P1 | Rising edge | | High level | Low level | Reverse |
| P1 | Falling edge | | Low level | High level | Reverse |
| P2 | Rising edge | High level | | Low level | Forward |
| P2 | Falling edge | Low level | | High level | Forward |
| P2 | Rising edge | Low level | | High level | Reverse |
| P2 | Falling edge | High level | | Low level | Reverse |
| P3 | Rising edge | Low level | High level | | Forward |
| P3 | Falling edge | High level | Low level | | Forward |
| P3 | Rising edge | High level | Low level | | Reverse |
| P3 | Falling edge | Low level | High level | | Reverse |

the amplitude of the receiving coil passing cross-point is vulnerable to antenna vibration and shake, causing the dither of signal amplitude. If it is impossible to make a corresponding adjustment by comparing thresholds, leak detection at cross-point will happen; if threshold is increased, the detection ability at the cross-point will be enhanced, but drifting will happen to its duty ratio, which will also lead to some deviation of the train position and speed.

The signal processing method based on phase is less dependent on the amplitude of the receiving signal, free from the situation of leak unit ring detection. At the expense of the increased pair number of coils, system liability has been sharply enhanced. Hence, the position and speed detection system for low-speed Maglev train will use the signal processing method based on phase.

## 6.5 System Design and Implementation

This section conducts engineering design for the position and speed detection system based on inductive loop table [13–16]. Firstly, software and hardware design for the position and speed detection system is introduced. Secondly, from the perspective of improving system anti-interference ability, a filtering circuit is designed and relevant parameters are determined. The power processing circuit is designed for the system's ground and vehicle-mounted actuating device of two sets of systems. Finally, experimental results are obtained. By operating test on 1.5 km test line, the design validity for the system structure, algorithm, and analysis can be verified.

## 6.5.1 Design of Analog Circuit

### 6.5.1.1 Band-Pass Filter Design

In order to effectively improve the circuit capacity of filtering and integration level and to reduce power consumption and temperature impact, this section selects chip of switched-capacitor filter to conduct filtering processing on the input signal.

The switched-capacitor filter is a large-scale integrated circuit filter, composed of MOS switch, MOS capacitors, and MOS operational amplifiers, with three remarkable characteristics:

1. Firstly, under the condition of a confirmed system clock, the device filter property only depends on the capacitance ratio. This capacitor is made by a certain process, of which the temperature characteristic is fairly stable and the ratio can achieve a very nice precision.
2. Secondly, under the condition of fixed circuit composition, input and output characteristic of the designed filter can be easily adjusted through changing working frequency.
3. Thirdly, digital filter generally needs to undergo analog/digital and digital/analog conversion. However, this filter can directly handle analog signal, further simplifying the circuit structure, and system stability can be promoted.

Figure 6.30a is the simplest switched-capacitor filter. When switch $k$ is in the left, input voltage source $u_i$ charges to capacitance $c_1$; when $k$ is in the right, capacitance $c_1$ discharges to voltage source $u_2$. When switch works with $f_c$ higher than signal frequency, making $c_1$ alternately connected between $u_1$ and $u_2$ two voltage nodes, average current $i = f_c c_1 (u_1 - u_2)$ formed by electric charge transferred by $c_1$ between $u_1$ and $u_2$ can be seen as a resistance with certain value connected in series between voltage source $u_1$ and $u_2$ in Fig. 6.30a, and the value of this equivalent resistance is $1/(c_1 f_c)$. Therefore, switched-capacitor circuit in Fig. 6.30a is equivalent to a first-order active low-pass filter, of which transfer function can be expressed as follows:

$$G(j\omega) = j \bullet \frac{c_1 f_c}{c_2 \omega} \tag{6.10}$$

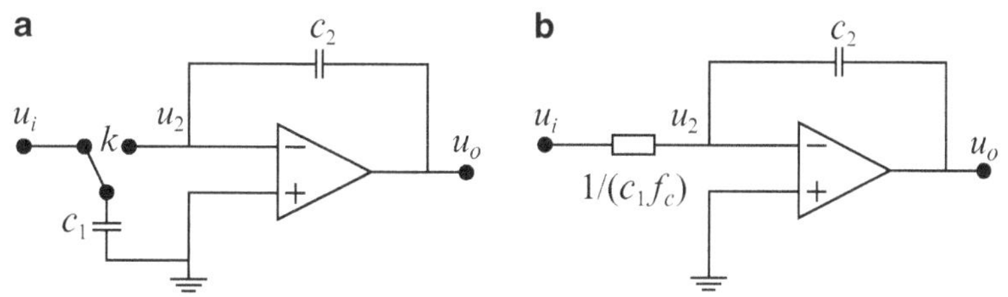


**Fig. 6.30** Switched-capacitor circuit and its equivalent circuit

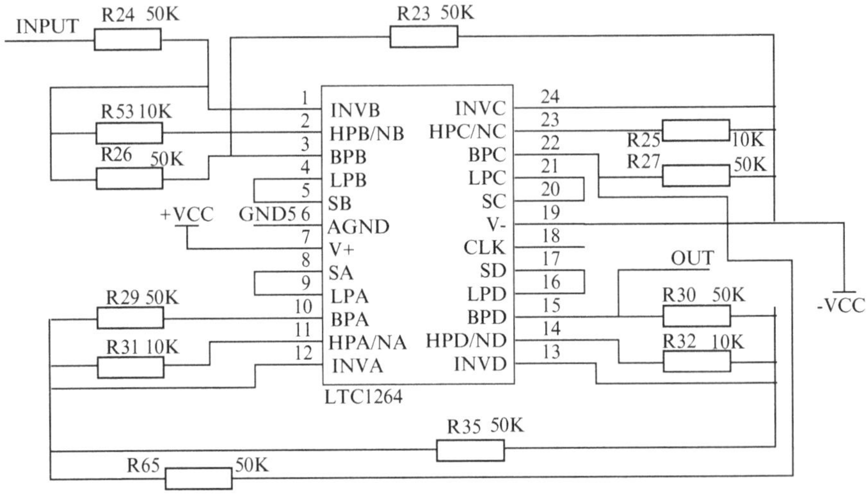


**Fig. 6.31** Band-pass filter

It can be seen from Formula (6.10) that the transfer characteristic of a switched-capacitor filter is decided by $c_1/c_2$ and switch frequency $f_c$. Since Fig. 6.30b is an integral circuit. In this way, the switched-capacitor filter can simulate corresponding circuit of the filter, further realizing *LC* filter, active filter, and other characteristics.

The position and speed detection system uses LTC1264 chip manufactured by Linear Technology Corporation to realize band-pass filtering as shown in Fig. 6.31. The chip includes four independent high-speed second-order switched-capacitor filters and the maximum programmable frequency is 250 kHz. The second-order low-pass, high-pass, band-pass, and band rejection filter can be realized by adding three to five resistances to each filter model. Its center frequency can be adjusted through external reference clock, and the relationship between center frequency $f_c$ and reference clock $f$ is $f/f_c = 20 : 1$. A maximum eight-order band-pass filter can be realized by connecting four band-pass filters in series. It can be seen from amplitude versus frequency curve in Fig. 6.32 that its -3 dB bandwidth is 10 kHz.

#### 6.5.1.2 Low-Pass Filter Design

Based on the introduction in Sect. 6.4, it can be known that the signal processing method based on phase receives signals transmitted by the railway inductive loop table through two matching coils whose distance is a half of unit ring length and obtains exclusive or square wave reflecting train relative position after shaping and exclusive or treatment of receiving signal. Under ideal condition, based on different relative position of two receiving coils, the phase of receiving signals is either in-phase or anti-phase, namely, phase difference is 0° or 180°. However, the coil

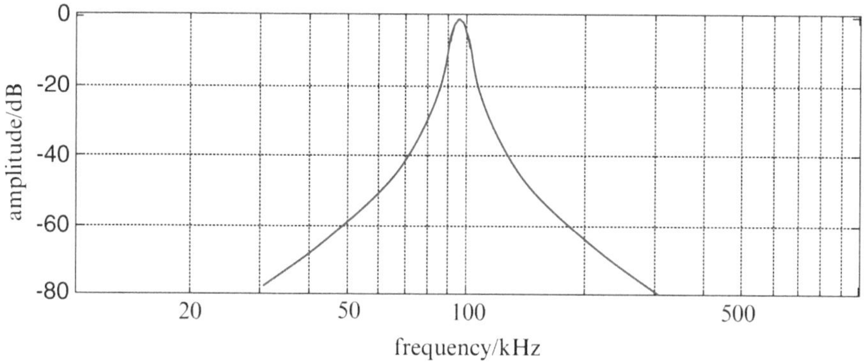


**Fig. 6.32** Amplitude-frequency characteristic of filter designed by Fig. 6.31

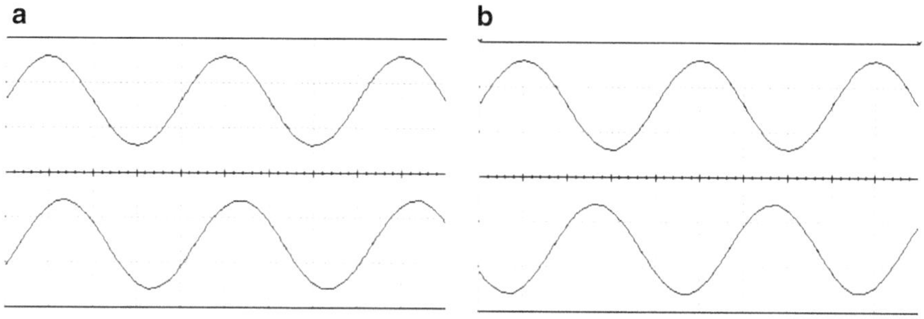


**Fig. 6.33** Oscilloscope screenshot of phase drifting signals output by two receiving coils: (**a**) in-phase drifting and (**b**) anti-phase drifting

parameter cannot be absolutely consistent, which will lead to the phase drifting existing in output signal after parallel resonance of two coils and a high-frequency narrow pulse existing after exclusive or treatment as shown in Fig. 6.33.

In order to solve the problem mentioned above, a low-pass filter shall be added after exclusive or treatment, and then the filter output shall be shaped. As a result, an accurate position square wave can be obtained (Fig. 6.34).

Considering that shaping treatment shall be done after the low-pass filtering, and according to the requirements on characteristics, for the flatness and group delay within designed filter, the passing band is not high. Chebyshev filter with good cutoff characteristic can be chosen. Figure 6.35 shows a normalized second-order Chebyshev low-pass filter with the passing band ripple of 0.01 dB, of which corresponding transfer function is listed as follows:

$$G(s) = \frac{1}{c_1 c_2 s^2 + 2c_2 s + 1} \tag{6.11}$$

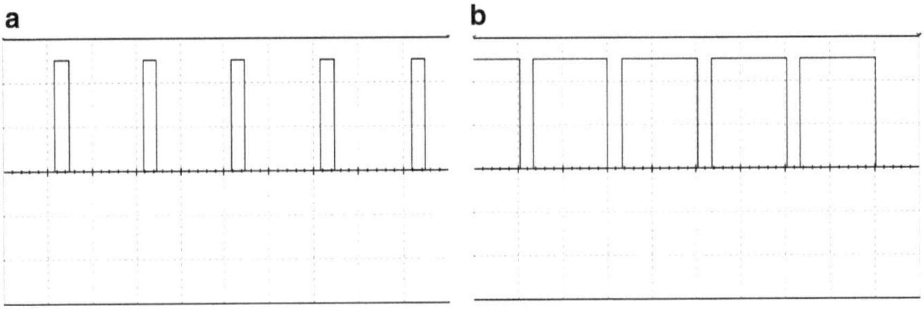


**Fig. 6.34** Oscilloscope screenshot of waveform shown in Fig. 6.33 after shaping and exclusive or treatment: (**a**) in-phase drifting and (**b**) anti-phase drifting

**Fig. 6.35** Normalized second-order Chebyshev filter

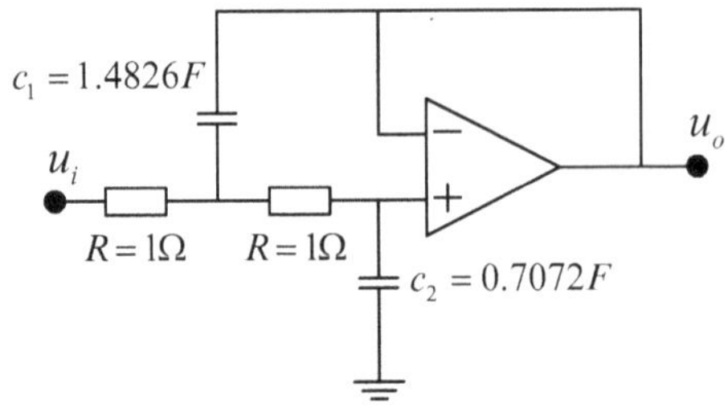


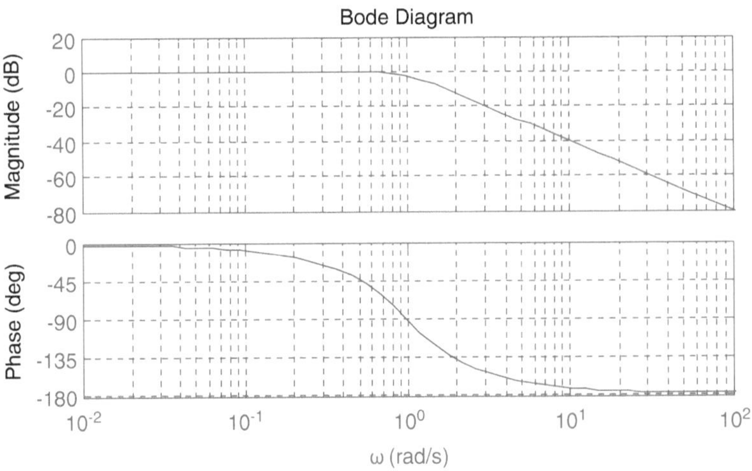


**Fig. 6.36** Bode diagram

Further, the amplitude-frequency and phase-frequency characteristic curve can be obtained as shown in Fig. 6.36.

Normalized transformation formula can be shown below:

$$R' = R \times Z \quad L' = \frac{L \times Z}{F} \quad C' = \frac{C}{F \times Z} \tag{6.12}$$

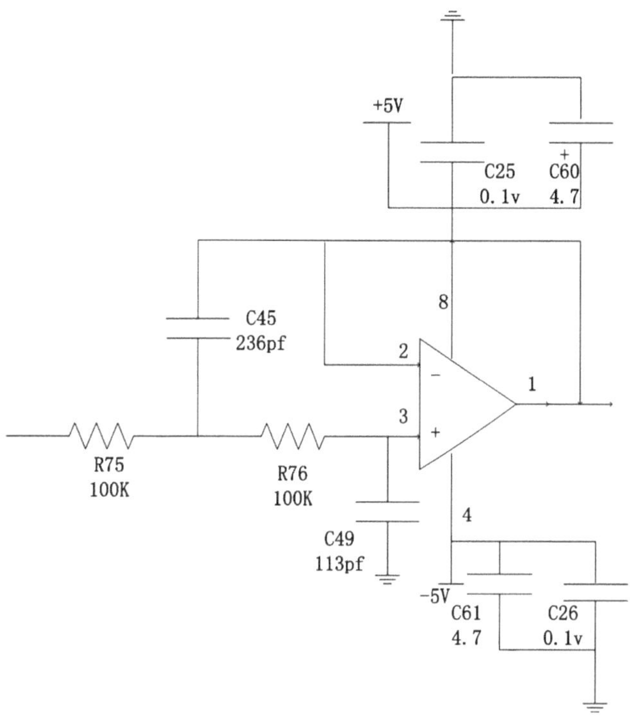


**Fig. 6.37** Second-order Chebyshev filter

Therein, $Z$ is the coefficient of impedance transformation, $R'$ is the numerical value after impedance change, and $F$ shows the ratio of given parameter frequency and corresponding known filter parameter frequency. The normalized cutoff frequency is $1/2\pi$ Hz, $F = f_c \times 2\pi$.

The actual cutoff frequency of filter is 10 kHz, $F = 6.28 \times 10^4$; the coefficient of impedance transformation is $Z = 10^4$; resistance and capacitance in corresponding Fig. 6.35 change into $R' = 10^4\Omega$, $c_1' \approx 236pf$, and $c_2' \approx 113pf$; and the actual circuit of filter obtained is shown in Fig. 6.37.

## 6.5.2 *Design of Digital Circuit*

The core processing module, TMS320LF28335 manufactured by TI Corporation, is used to realize relative position detection, speed calculation, and direction discretion [17, 18]. Its external clock frequency is 30 MHz, and the maximum internal clock frequency (150 kHz) can be realized through the internal frequency de-multiplication, featured by 16-bit addresses and data bus and supported for external expansion, integrating 16 analog-digital conversion channels with 88 programmable digital I/O interfaces.

**Fig. 6.38** DSP program flow

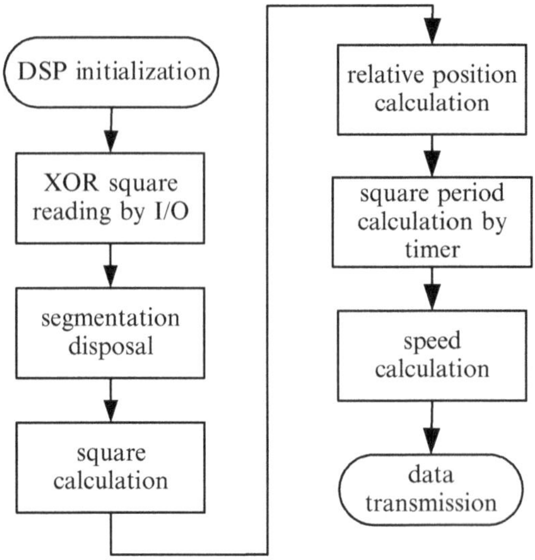


DSP program loaded circuit: TMS320LF28335 can guide the program from internal FLASH or from external memory. The main program flow is shown in Fig. 6.38.

Protocol processing module: The programmable logical device uses EP1C6T144C8 chip manufactured by ALTERA Corporation to realize data communication between the position and speed detection system and other system.

### *6.5.3 Design of Power Drive*

Both the vehicle-mounted position and speed detection system and the ground position and speed detection system adopt the same power drive, which separately drives the railway inductive loop cable and vehicle-mounted transmitting antenna [19]. The drive uses a high-voltage floating MOS grid drive integrated circuit IR2110 as bridge driver chip, which can directly drive high-level and low-level field effect transistor, making H-bridge switching circuit simplified and improving circuit performance. In the power amplifier circuit, power switching elements generally adopt isolated drivers. Since chip IR2110 is featured by both opto-isolator and electromagnetic isolator, small in size and free from expansion, it is a driven integrated chip with relatively good performance, which can be directly used in power amplifier, leading to a more compact design. Hence, this chip is chosen as the driven control chip for power amplifier. As IR2110 does not possess the function of the input signal adjustment, CPLD can be used to constitute signal generator for its driven preamplifier, and driving system adopts EMP3032A manufactured by ALTERA Corporation as square wave output chip.

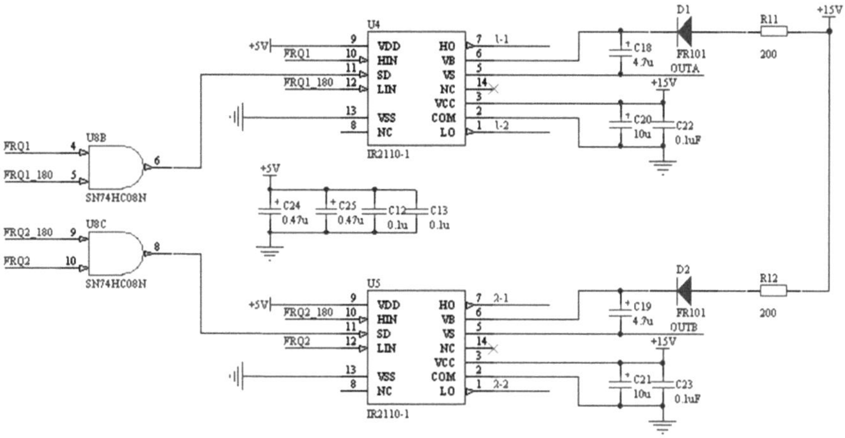


**Fig. 6.39** *H*-bridge drive

To guarantee that the signal is strong enough, drive current shall be generally maintained above 0.8 *A*, and the power amplification is needed. After supply voltage has been confirmed, outputting power as high as possible and improving the transfer efficiency of this circuit are main issues for the power-driven circuit research. Hence, two IR2110 are used to consist of *H*-bridge drive circuit as shown in Figs. 6.39 and 6.40. In order to ensure enough drive current and enough large voltage, OCL adopts the transformer coupling mode and transformer boost mode under given load condition [20, 21].

### *6.5.4 Test of Position and Speed Detection of Low-Speed Maglev Train*

1. Test of position and speed detection with maximum speed to 20 km/h

It can be seen from Fig. 6.41a that the vehicle-mounted position and speed detection system can accurately generate three exclusive or square waves and position square wave with precision reaching 10 cm after segmentation treatment reflecting train relative position. Figure 6.41b shows the speed of the moving train calculated according to segmentation square wave, and it can also be seen from speed curve that maximum speed of moving train is 20 km/h.

2. Test of position and speed detection with maximum speed to 65 km/h

Figure 6.42a shows that when train starts statically and accelerates to 65 km/h, and then decelerates to 35 km/h, three exclusive square waves and segmentation square wave generated by vehicle-mounted position and speed detection system are

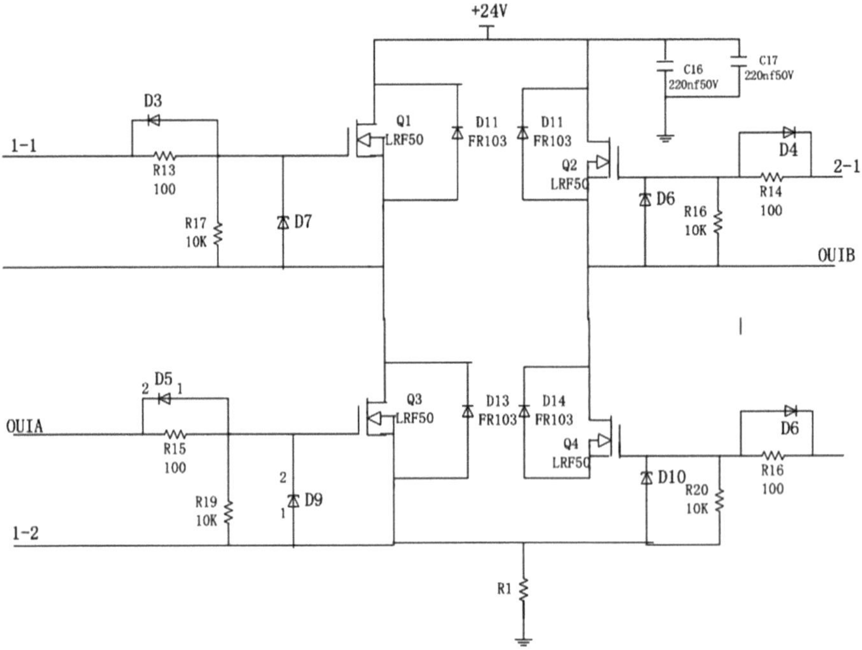


**Fig. 6.40** H-bridge circuit

also free from leak detection of the unit ring. Figure 6.42b shows the speed of the moving train calculated by segmentation square wave, and it can also be seen from speed curve that maximum speed of movingtrain is 65 km/h.

## 6.6 Conclusion

This chapter introduces the position and speed detection technology based on the inductive loop cable. Firstly, space magnetic field generated by the inductive loop cable is investigated; based on the properly simplified railway inductive loop cable, the distribution of magnetic flux density is respectively calculated in $x$, $y$, and $z$ directions; secondly, through comparing the quality factor of different coil types, the basic structure of vehicle-mounted receiving coil is designed; thirdly, based on the analysis of signal processing methods of the vehicle-mounted position and speed detection system aiming at the existing cumulative error problem due to leak unit ring detection of the signal processing method based on amplitude detection, signal processing method based on phase detection is designed, and its operating principle is also analyzed. By comparing with previous methods, the signal processing method based on phase detection is used. Finally, the engineering of the position and speed detection technology for low-speed Maglev train is realized in practical engineering.

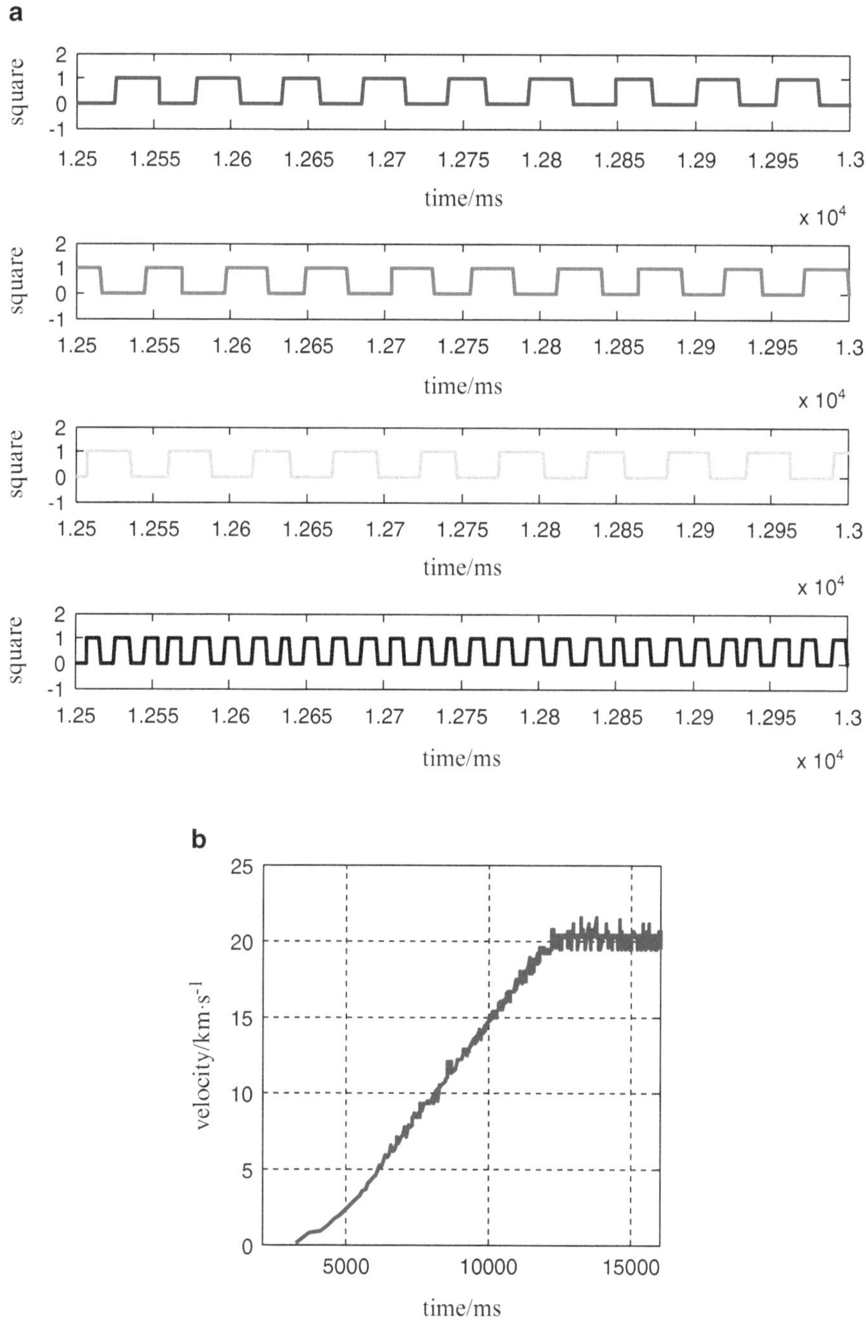


**Fig. 6.41** Position square wave and speed curve with accelerated speed to 20 km/h. (**a**) Three exclusive or square waves and square wave after segmentation treatment. (**b**) Speed curve

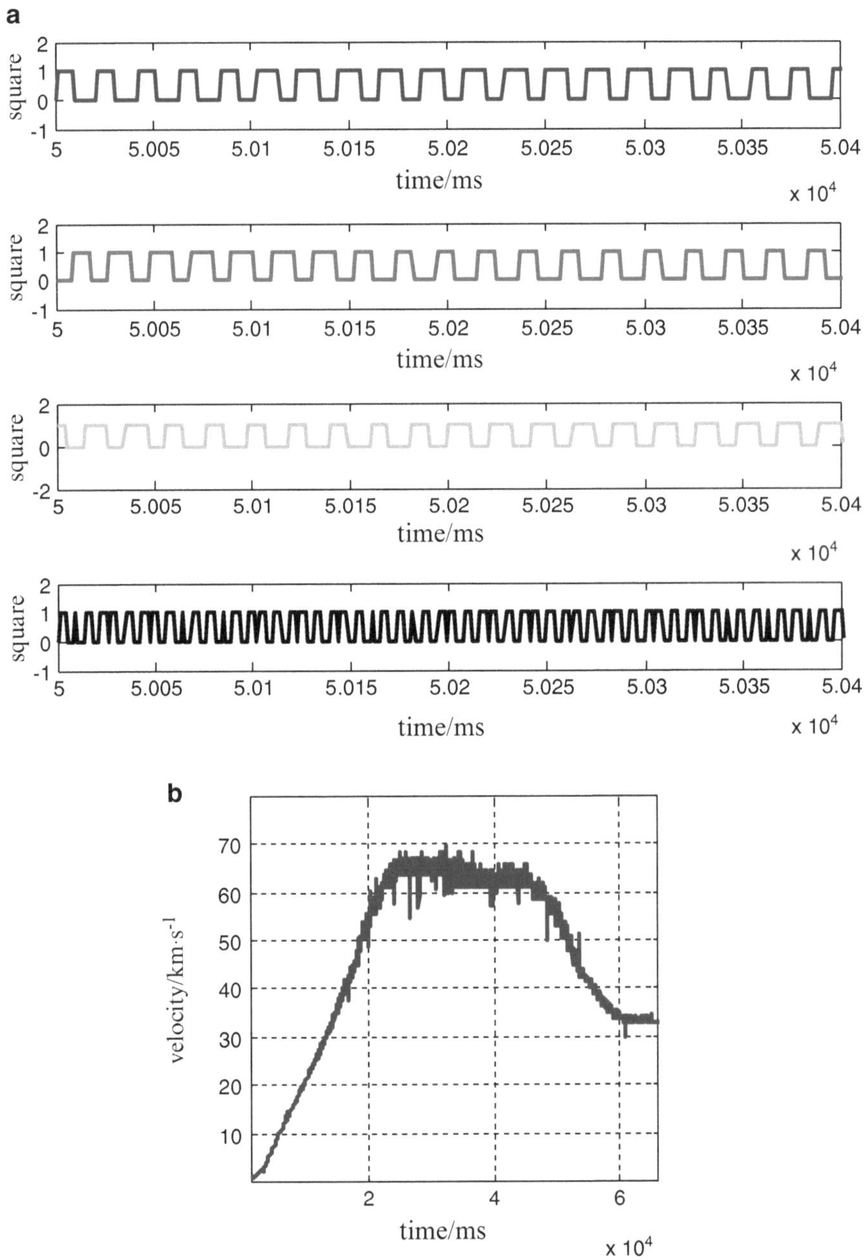


**Fig. 6.42** Relative position square wave and speed curve with maximum speed to 65 km/h. (**a**) Three exclusive or square waves and square wave after segmentation treatment. (**b**) Speed curve

# References


1. Wu Xiangming. Maglev train [M]. Shanghai: Shanghai Scientific and Technical Publishers; 2003.
2. Atzpodien, H. -C. Transrapid Maglev system - fields of application [C]. In: MAGLEV'2006: the 19th international conference on Magnetically Levitated Systems and Linear Drives, 13-15 September 2006, Dresden, Germany; 2006.
3. Tejima Y, Zhirong Liu. Commercialization of HSST, an access line for the 2005 world exposition in Aichi, Japan [J]. Convert Technol Electric Traction. 2005;3(1):50–2.
4. Syunzo ISHIMOTO, Masato KATO. The first urban Maglev transport application in Japan [C]. Maglev 2004 proceedings. Shanghai;, China, October 26–28, 2004. 1: p. 298–306.
5. Shin BC, Kim WJ, Park DY, Beak JG, Kang HS. Progress of urban Maglev program in Korea [C]. The 9th World Congress Railway Research, Lille, France, May 22–26, 2011.
6. Xianglei Song, Fengshan Dou, Chunhui Dai, Zhiqiang Long. Modeling and design of the speed and location system for low speed Maglev vehicle [C]. International conference on intelligent computation technology and automation. 2012. 1:635–41.
7. Chunhui Dai, Xianglei Song, Fengshan Doû, Zhiqiang Long. Study on the speed and location system for low speed Maglev vehicle [C]. IEEE international conference on automation and logistics. 2011. 1:635–41.
8. Chunhui Dai, Zhiqiang Long, Yunde Xie, Song Xue. Research on the filtering algorithm in speed and position detection of Maglev trains [J]. Sensors. 2011;11(7):7204–18.
9. Chun Dai, Fengshan Dou, Xianglei Song, Zhiqiang Long. Analysis and design of a speed and position system for Maglev vehicles [J]. Sensors. 2012;12(7):8526–43.
10. Yang Jianyong, Lian Jisan. Positioning and speed-measuring of maglev vehicle based on rail cable [J]. Electr Drive Locomot. 2001;1:30–2.
11. Sun Zhang. The low-speed maglev train pilot system through pilot reviews in Changsha [J]. Urban Mass Transit. 2001;41(4):69–73.
12. Liu Zhiming, Wang Yongning. Low-speed Maglev train technology and engineering research and development of our country [J]. Rail Transit. 2006;2:46–50.
13. Tan Qiquan. Theory of a new radio induction system and its application [D]. PhD dissertation. Chengdu: Southwest Jiaotong University Doctor Degree; 2003. p. 6.
14. Tan Qiquan, Lian Jisan. Application of new wireless induction technology to location of rail moving objects [J]. Electr Drive Locomot. 2003;1:12–4.
15. Chen Jiaxiang. Induction wireless technology [J]. Electr Autom. 2000;22(2):54–6.
16. Cheng Wangbin, Chen Jin, Chen Xin. Application of new radio induction technology to position detection [J]. J Electr Meas Instrum. 2010;24(4):379–83.
17. Chen Jin. The department of computer and information engineering [J]. J Electr Meas Instrum. 2009;23(1):58–63.
18. Yang Jianyong. Maglev train automatic control (ATC) system [D]. PhD dissertation. Chengdu: Southwest Jiaotong University Doctor Degree; 2000. p. 6.
19. Li Hua. Optimization of cross loop with long distance transmission [J]. Railw Signal Commun Eng. 2011;8:73–5.
20. Williams A, Taylor FJ. Electronic filter design [M]. Beijing: Science Press; 2008.
21. Ronger S. LC filter design and production [M]. Beijing: Science Press; 2006.

MIX
Papier aus verantwortungsvollen Quellen
Paper from responsible sources
FSC® C105338
FSC
www.fsc.org

If you have any concerns about our products,
you can contact us on
**ProductSafety@springernature.com**

In case Publisher is established outside the EU,
the EU authorized representative is:
**Springer Nature Customer Service Center GmbH**
**Europaplatz 3, 69115 Heidelberg, Germany**

Printed by Libri Plureos GmbH
in Hamburg, Germany